T0212304

DEPOSITION OF
THE SEDIMENTARY ROCKS

DEPOSITION OF
THE SEDIMENTARY ROCKS

BY

J. E. MARR, Sc.D., F.R.S.

*Woodwardian Professor of Geology
in the University of
Cambridge*

CAMBRIDGE
AT THE UNIVERSITY PRESS
1929

CAMBRIDGE
UNIVERSITY PRESS

University Printing House, Cambridge CB2 8BS, United Kingdom

Cambridge University Press is part of the University of Cambridge.

It furthers the University's mission by disseminating knowledge in the pursuit of education, learning and research at the highest international levels of excellence.

www.cambridge.org
Information on this title: www.cambridge.org/9781107492530

First published 1929
First paperback edition 2015

A catalogue record for this publication is available from the British Library

ISBN 978-1-107-49253-0 Paperback

PREFACE

I HAVE in this book endeavoured to give a general account of the conditions which have controlled the distribution of the various kinds of sedimentary deposits in time and space, taking into account also the features relating to their organic contents which are necessary to my purpose. I have avoided overloading the book with excess of detail. There are already many treatises devoted to the characters of the deposits. A succinct account will be found in Hatch and Rastall's *Petrology of the Sedimentary Rocks*, while a very full description has lately been published in the American *Treatise on Sedimentation* by W. H. Twenhofel. A large number of instances cited in illustration of various points are derived from the Lower Palæozoic rocks simply because I am more acquainted with these, as the result of actual study, than with those of later date.

It is my pleasant duty to acknowledge assistance received from many friends, especially among my own colleagues in the Sedgwick Museum; to H. Woods, F.R.S., and Dr G. L. Elles I owe help given in various ways; W. B. R. King has given me special assistance in many matters, including the drawing of the diagrams. I am above all indebted to Dr R. H. Rastall, without whose help the book could not have appeared, as it was written under disadvantages which but for him would have been insuperable. He copied out the whole MS. for me and at the same time criticised it and made most valuable suggestions: he has further increased my indebtedness by revising the proofs and compiling the index.

In addition to help from my colleagues I have had aid from F. A. Potts, M.A., while J. A. Steers, M.A., and W. F. Whittard, Ph.D., have furnished me with material which is acknowledged in the text.

J. E. M.

Cambridge
May 1929

CONTENTS

CHAPTER I

INTRODUCTION: CHRONOLOGY AND PHYSICAL CONDITIONS

PHYSICAL Geography is the geology of the present day, and conversely geology is the sum of the physical geographies of the past and present.

In attempting to restore the history of the Earth it is therefore necessary to establish a chronology, and having done so, to make a study of the rocks of each of the periods so established, in order to discover what events were happening at each particular period, with a view to elucidating the physical geography of that period. This is mainly, though not exclusively, done by a study of the sediments formed during the period: not exclusively, for the igneous rocks then formed must also be taken into account, and the various changes generally included under the head of tectonics. As a matter of fact the establishment of chronology and the study of the evidence for the conditions of deposit have to a great extent been carried on simultaneously.

A study of the sediments then is of prime importance to the geologist, and great progress has been made in it, but much still remains to be done. Not only must the petrographical characters of the sediments be taken into consideration, but other matters, especially their areal distribution and their organic contents.

A good deal of information has been acquired on all these points, but particular attention has been paid to the petrology. Even here the progress has not been so great as is desirable. The attention of geologists has it is true, during the last half century, been devoted in an exceptional degree to petrology, but the fascinating problems connected with the igneous and metamorphic rocks have to some extent diverted attention from sedimentary petrography, though this study has by no means been ignored; Sorby himself made most valuable contributions to it, especially in his two Presidential Addresses to the Geological Society in 1879 and 1880.

While much attention has been devoted to ancient sediments, those now being formed have been by no means neglected. Great impetus was given to this branch when the careful examination of the sea-floor, as the result of deep soundings, was instituted, and we now possess an extensive literature bearing upon the matter, prominent in which is the description of the sediments, by Sir John Murray and the Abbé Renard in the *Challenger* Reports[1].

Even here, however, the fascination of the deposits of the abyssal portions of the oceans tended to check interest in the more prosaic deposits of the shallows, which are nevertheless of particular importance to the geologist.

It behoves us then to make further study of ancient and modern sediments alike, to a much greater extent

[1] Report of the Scientific Results of the Voyage of H.M.S. *Challenger*: Deep Sea Deposits, 1891.

than has hitherto been done. That the present is the key to the past is in the main true, but there are cases when the past is the key to the present, as has been illustrated on many occasions during the history of geology, and notably so in this particular research into the nature and causes of formation of the sediments.

The student of ancient sediments has two important advantages over the oceanographer examining those now being formed. In the first place, whereas the last named, save between tide-marks, is dependent upon soundings, often made at considerable distances apart, the former can observe continuous sections often over wide tracts of country. Secondly, whereas the oceanographer can only regard the superficies, the geologist is able to deal with his material in three dimensions. On the other hand, when dealing with modern deposits we know the exact type of area in which they are formed; as estuary, bay, gulf and so forth, whereas in the case of the ancient deposits, this is a matter of inference. Only by continued study of ancient and modern deposits are we likely to make much progress.

The establishment of a chronology was begun when geology was in its infancy. Without such a time-scale it is obvious that little progress could be made in the prosecution of the science. In establishing a time-scale two main tests have been applied, namely that of superposition and that of the sequence of organisms. The former necessarily came first, for the sequence of organisms could only be primarily established in the case of a known upward succession of undisturbed strata. That such regular superposition does occur was realised among others by John Woodward, as set forth by him in his *Essay toward a Natural History of the Earth*, published in 1695, though the cause was incorrectly explained and the time taken for the accumulation of the sediments was greatly underestimated.

The importance of the included organisms was first clearly set forth by William Smith in several publications, especially in his *Strata Identified by their Organized Fossils*. As the result of the recognition of such a sequence, not only could the order of the strata be detected in any given area, but a correlation of the rocks of widely distant regions could also be made. This is a subject which has been much debated, and will be considered more fully in a later chapter. Though the utility of these organisms as aids to chronology is now fully recognised, lithological character must not be altogether ignored in this connection, being distinctly useful as a subsidiary aid.

In the second part of the enquiry, which deals with the conditions that prevailed during the formation of the deposits of any particular period, the lithology of the sediments is obviously of primary importance, though very valuable information is also afforded by a study of the included organisms. From this point of view the organisms have not been studied to anything like the extent which is evidently required. In the great task of ascertaining chronological sequences attention has naturally been devoted mainly to similarities of the faunas and floras in deposits of the same age, with the object of correlation, and this has often been done without paying much attention to the differences, save in a general way. The importance of the last named is obvious, and now that we have established a time-scale successfully applicable over wide areas, it is time that attention should be directed more particularly to differences in the distribution of the organisms when traced laterally.

The study of the sediments and of their organic contents, with a view to the establishment of chronologies and to the elucidation of the physical conditions prevalent at the different geological periods, will be pursued in the following chapters, taking the evidence of lithology and included organisms on, first, chronology, and second, conditions of deposition.

It may be well to devote a few words to the use of the term *sediment*. The classification of rocks now in use is to some extent arbitrary. We are accustomed to speak of two great classes, the Igneous, and the Aqueous, Stratified or Sedimentary. There is some objection to the use of each one of the three latter terms. To the term sediment it may be objected that some of the accumulations placed in this group are not sediments, as for instance surface soils: the objection is valid, but as the term has got into general use it may well be retained, as no other has been proposed which is not open to objection. The terms stratified and aqueous are also not sufficiently comprehensive. The word epiclastic has a more definite meaning than sedimentary, but unfortunately does not include all the rocks with which we are concerned, for volcanic ashes belonging to the pyroclastic group directly concern us, forming, as they do sometimes, definite marine sediments with organisms, differing in no way from non-volcanic sediments with which they are associated save in the source of supply of their material.

The word sediment then will be used hereafter in its usual and generally understood sense, to include the various accumulations, whether terrestrial or sub-aqueous in origin, other than lava-flows and intrusive igneous rocks, which constitute the geological column.

CHAPTER II

CHRONOLOGY

WHEN establishing the succession of strata it is necessary to make subdivisions, for without these no correlation of deposits of the same age in different areas can be attempted. Such a classification might be purely arbitrary, as by division of strata into groups of equal thickness, but this would obviously be of little practical value, and it is necessary to attempt a classification which is so far as possible natural.

A classification supposedly natural, into Primary, Secondary and Tertiary, was long ago suggested. It was founded on insufficient data and is known to be faulty; but the present admittedly imperfect classification has gradually been evolved from it. When the sediments were first divided into Primary, Secondary and Tertiary it was believed that violent changes occurred in the intervals between these periods, causing physical catastrophes and complete extinction of life, new floras and faunas being created at the beginning of the ensuing period. Similar changes on a smaller scale were believed to have occurred also in the intervals between the principal divisions, or systems, into which these three periods were further split up. Such a classification was then held to be natural, and in most cases world-wide.

In the present state of our knowledge it is recognised that even the larger divisions established in this way by the occurrence of breaks cannot be of world-wide application, for it is obvious that the whole earth cannot be covered by land (or by sea) at the same time, and it has been abundantly proved that continuous sedimentation was proceeding in some areas while the breaks were being produced in others. Nevertheless the divisions founded on breaks are far from arbitrary, and such breaks have evidently occurred frequently and over wide areas.

As the oceans are mainly areas of deposition and the lands of erosion, the greater part of the sediments preserved are marine. A cycle of occupation of any region by the sea between two terrestrial periods will include three phases. Beginning with submergence of a pre-existing land, the early stage of a marine cycle will be marked by shallow-water deposits: the middle period by more open-water beds; and the final stage by the recurrence of shallow-water deposits. Each of these will be marked by certain characteristic features of the sediments, and the whole will usually be complicated by epicycles of minor emergences and submergences during the main cycle.

Towards and at the close of the cycle the sediments then formed will gradually emerge to become part of the succeeding land, and consequently will undergo erosion. It results from this that although the deposits formed during the lower and middle phases will probably escape, those of the upper will be partly if not entirely removed, and the earliest deposits of the succeeding marine period will be laid unconformably upon them. An important discontinuity thus produced will probably be accompanied by a marked palæontological break, due to more than one cause.

Fig. 1 *A* represents a section of a series of conformable strata *a–k*, the vertical lines being intended to represent the ranges in time of a number of fossils. In Fig. 1 *B* an attempt is made to show what happens

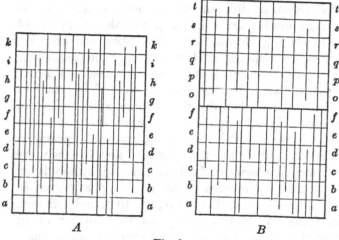

Fig. 1

when there has been a physical break, resulting in the denudation of the beds *ghik*, and the deposition of another set *op*... unconformably upon those deposits of the earlier set which remain. As the result of this we note, first, that the relics of organisms which existed in the area during the deposition of *ghik*, and were entombed in those strata, are destroyed by the processes of denudation, and a large number of organisms which lived long after the deposition of f, and disappeared not simultaneously but at different times during the period when denudation was in operation, seem to become extinct simultaneously at the top of f, though if we could visit an area which was receiving sediment during the period of denudation, we should find them dying out in the rocks of that region at different levels. Furthermore, whilst denudation is going on, a longer or shorter period of time elapses, during which the upheaved area receives no deposit, and accordingly no organisms which lived during that period are preserved in the upheaved area. During this time a set of deposits lmn may have been laid down elsewhere, and besides the gradual disappearance of some of the organisms $ab...k$, there will have been a gradual appearance of new species. When the upheaved area is once more submerged, a new set of deposits $op...$ is accumulated in it, and the species which gradually appeared in adjoining regions will now migrate to it, and will seem to come in simultaneously at the bottom of o; accordingly we may find that there is not a single species which passes through from f to o, and the palæontological break in this area is complete, though it is clear that it only implies local change, and that we may and indeed must find intermediate forms in other regions.

Owing to the occurrence of these breaks separating deposits formed during periods intervening between those during which the breaks were produced, we have a means of making a chronological classification, founded partly on physical, partly on organic changes, and such was the classification put forward when the major divisions or systems were established. It is not of world-wide application, but as it is now largely held that the changes which converted land into sea and vice versa were perhaps in some degree world-wide, such a classification is probably much more natural over wide areas than has been supposed. The advance of our knowledge has enabled us to use this method with greater refinement than formerly. The geologists of the United States have of late years strongly advocated its application[1].

As the great systems now generally adopted were established in Britain, the classification in use is to a very high degree natural in this country, but even here changes have had to be made. For example, it is now known that the Cenomanian transgression is a more important break, so far as N.W. Europe is concerned, than that separating the Jurassic and Cretaceous systems.

A different method based on peculiarities of lithology has been used in defining some systems, as for example the New Red Sandstone, whose special features are the result of deposition under arid continental conditions.

Furthermore, an entirely new scheme was adopted in delimiting the major divisions of the Tertiary rocks. In Tertiary times an appreciable number of still living species had come into existence, and the

[1] See Chamberlin, T. C., "Diastrophism as the Ultimate Basis of Correlation" in *Outlines of Geologic History*, Chicago, 1910.

division into Eocene, Miocene and Pliocene was founded on the percentages of living and extinct forms in each. The plan has proved to be largely unworkable, owing to the difficulty caused by the splitting up of species. For example, if what was originally supposed to be one species, and that a living one, has been subsequently divided into half a dozen, of which five are extinct, the percentage of extinct forms in the later lists is obviously greater than in the earlier ones. This vitiates the method, and although the classification was founded upon this principle, its recent development has been due to methods analogous to those employed in the case of the pre-Tertiary rocks. From these remarks it will be gathered that although physical breaks have been utilised with much success in determining the larger divisions or systems, the method pursued has not been uniform, and modifications in the classification are still being made; this is specially the case in areas remote from those in which the systems were originally defined.

Many of the systems have geographical names, as Cambrian, Jurassic: others like Carboniferous and Cretaceous have a lithological significance, while, as above noted, the names of the Tertiary divisions, as Eocene and Miocene, refer to the percentage of living and extinct forms in each. As our knowledge advances we find as the result of experience that a classification founded on the changes in organisms is more useful and more widely applicable than one founded on other considerations, not only in the case of the Systems, but also with the smaller divisions, the Series and Stages, marking episodes of varying degrees of magnitude in a great marine cycle, and we may confidently anticipate a general classification in the future, founded entirely on palæontological considerations. As a matter of fact such a classification for the Lower Palæozoic rocks was long ago adopted by N. P. Angelin, in his *Palæontologia Scandinavica*, published in 1851, a new edition of which appeared in 1878.

The classification founded on trilobites, as given in the 1878 edition, is as follows:

Regio VIII.	Cryptonymorum (Encrinurorum)	$= E$
VII.	Harparum	$= DE$
VI.	Trinucleorum	$= D$
V.	Asaphorum	$= C$
IV.	Ceratopygarum	$= BC$
III.	Conocorypharum	$= B$
II.	Olenorum	$= A$
I.	Fucoidarum	

The Fucoid division is now known to correspond with that containing the *Olenellus* fauna, and may be spoken of as Regio Olenellorum. The Regiones Olenorum and Conocorypharum were placed in the wrong order by Angelin, otherwise the classification is correct. The letters *BC* of the Regio Ceratopygarum indicate that this is a passage fauna, and it is now known to be such between the Cambrian and

Ordovician, occurring on the horizon of the Trema-
docian. Similarly the Regio *DE* contains a passage
fauna between Ordovician and Silurian rocks, being
partly Ashgillian and partly Valentian. Thus there
are three Regiones in the Cambrian, those of *Olenellus*,
Conocoryphe (*Paradoxides*) and *Olenus*; the passage
fauna with *Ceratopyge*; two in the Ordovician, *As-
aphus* below and *Trinucleus* above; the passage fauna
between Ordovician and Silurian with *Harpes*; and
the one Silurian Regio with *Encrinurus*. It is obvious
however that many of the genera adopted are not
confined to the Regiones named after them. Apart
therefore from the passage faunas we have three
Cambrian divisions, two Ordovician and one Silurian,
and this is probably an approximate indication of the
true values of the systems as marking periods of time.

The classification shows rare foresight on Angelin's
part and may be regarded as an anticipation of the
type of classification which will ultimately be adopted
for all the fossiliferous sediments, and will probably
be applied to the Lower Palæozoic rocks with little
modification when that classification is made.

The establishment of breaks due to Revolutions
causing alternations in any given area of periods of
widespread oceanic conditions with sedimentation
and land conditions with erosion, has as stated been
useful for establishing major divisions in a chronological scheme of classification, but additional methods
are required for more detailed divisions founded on a
study of the lithology and especially of the organic
contents of the strata.

It has been said that the only way of actually proving contemporaneity of strata in distant areas is to trace them through the intervening tracts and so to construct stratigraphical maps, but this is not the case. Three kinds of stratigraphical maps may be constructed, one with lines drawn to separate sediments of different lithological character, such as limestone and mudstone; another will show lines separating sediments containing different faunas, and a third, lines indicating time-sequences, a line being drawn between the sediments of a period A and another B. Most of our present stratigraphical maps are ostensibly designed to be of the last type and are so to a great extent, but by no means exclusively: for instance, the line drawn at the base of the Millstone Grit is often a lithological and not a time-line. In actual fact the three kinds of lines are often largely concurrent, and when not so, it has been found as the result of experience that lines drawn by consideration of the faunas often approach more closely to the time-lines than those drawn by reference to lithological change. This will be considered more fully in the next chapter. Lithological changes are however by no means negligible in making correlations, and the rest of this chapter will be devoted to this matter, reserving the consideration of fossils until the following one.

LITHOLOGY AS A MEANS OF CORRELATION

The events which have happened in the past leave their impress on all succeeding periods, so that the characters of the sediments formed at any particular time will be dependent not only on the conditions then prevalent, but also to some extent on those of former periods. Thus for example the metamorphism of the rocks of one period will affect the physical and mineralogical characters of the sediments derived from them at a later period. Sediments of closely similar type have been formed again and again, at very different epochs, as for instance glauconitic sandstones as far back as the Cambrian, and in a general way any deposit now in process of formation can be matched among the strata of the past, but there are differences, especially as regards relative proportion, when areal distribution is taken into account. Consequently we find that prevalence of certain kinds of sediments over wide areas is characteristic of definite periods, though similar sediments have been formed in less degree at other times: as examples we may cite the widespread occurrence of the dark carbonaceous shales with graptolites in the Lower Palæozoic rocks and of coal-bearing sandstones in the period comprising the Upper Carboniferous and Permo-Carboniferous.

The above considerations will only aid correlation in a very general way on broad lines, but more detailed correlation is required and is possible to a considerable extent on lithological evidence. The use of sediments for purposes of correlation depends on their identity of character or of some of their characters over wide areas. Sediments are deposited in broad belts subparallel to coast-lines, of which more anon. They vary on the whole, when traced laterally, most rapidly in the belt nearest the coast-line, and become more uniform over wide areas in the belts away from the coast. Accordingly we find the deposits of the broad mud belt some distance away from the coast and the sediments formed of organic deposits outside that mud belt of the greatest utility for our purpose. Examples of the former are furnished by the three great mud belts of the Jurassic system— the Lias, Oxford Clay and Kimeridge Clay, and of the latter by the Chalk. The persistence of the Lias and Oxford Clay along their outcrop in England from north to south, and indeed far away on the Continent also, contrasts markedly with the varying nature and the limited distribution in space of each type in the case of the Lower Oolites between them, which were deposited in waters nearer to the coast-line.

It is only as the result of experience that we can ascertain what characters are useful for purposes of correlation over and above the main characteristics of a sediment. Whether a rock be sandstone, mudstone or limestone is of high importance, but many minor characters must be taken into account, including the presence of various extraneous matters, such as detrital mica, glauconite or phosphatic nodules, to mention only a few. The Penrith Sandstone, of Permian date, differs from the Triassic St Bees

Sandstone in the absence of detrital mica from the former and its abundance in the latter. Both are red sandstones of generally similar aspect, though there are other differences besides the one mentioned.

In some cases correlation is aided by characters actually impressed on the rocks subsequently to their formation. The Mountain Limestone of the British Carboniferous rocks was probably at the time of its formation very similar to what the Chalk now is. Subsequent changes have altered the former limestone to so great an extent that it presents marked differences from the Chalk, and even a small hand-specimen of the one can readily be distinguished from the other. Nevertheless the normal Mountain Limestone has an individuality of its own.

An interesting case of a widespread characteristic due to a structure formed subsequently to the deposit of the containing rock is furnished by the Wenlock Shale of Britain and several parts of the European continent. The rock is a mudstone of generally uniform character, though varying in colour from grey to black, but a noteworthy feature is the occurrence of large elliptical concretionary nodules: these of course are found in other rocks, but their wide prevalence in the Wenlock Shale renders them useful in helping the recognition of this deposit.

An even more striking example is supplied by the mudstones of Middle Cambrian age, known as the Menevian beds. They are characterised in North and South Wales and in the Welsh Borderland by a very marked rhomboidal jointing, which renders them easy to identify, and in one case, near Nuneaton, they were actually so identified in an area where their presence was not previously known. Here again we are dealing with a structure produced in the rock after its deposition. There must have been, in the case of the Wenlock Shale and of the Menevian beds special features of the rocks which caused the subsequent formation of the structures noted.

It will be seen then that experience determines the relative value of different lithological features for purposes of correlation. Much more work will be required in this matter. We are, for instance, still in doubt as to how far a study of the distribution of heavy minerals in sediments will aid us in our enquiry, but that their distribution may give much help in ascertaining the sources of supply of the sediment has been made clear by many observers, and recent work has already proved their value as aids to the identification and correlation of strata.

CHAPTER III

CORRELATION: THE USE OF FOSSILS

Fossils for Purposes of Correlation: Zones established as the Result of Experience

WILLIAM SMITH established his principle of correlation of strata by the use of fossils as a result of observation quite apart from any theoretical considerations, such as might be based on application of a knowledge of the lines of evolution. He was able to show that in a limited area, namely England, such correlation could be successfully established and was applicable in considerable detail. Smith's method was subsequently found to be applicable over much wider areas, and, as time went on, subdivisions were made in greater detail. In the course of years the use of fossils for this purpose became more and more extended, and at the present day no one doubts the utility of the method, though opinions differ as to how far it can be successfully extended in space and time. While admitting its utility for restricted regions, objection was raised to its use over very wide areas. Among the main objectors were Whewell, Huxley, and Herbert Spencer. They argued that so far from identity of fossils in two sets of strata in distant regions indicating their contemporaneity, it showed the contrary, for a considerable lapse of time would be needed for their migration from one area to another: accordingly, though similar sequences of faunas might be found in distant regions, each fauna would appear later in one area than in another. For such similarity of sequences Huxley proposed the term *Homotaxis*, maintaining that the homotaxial forms were not contemporaneous, each to each. Support seemed to be afforded to this view by Barrande's famous theory of "Colonies," applied to certain anomalies in the sequence of the faunas of the Lower Palæozoic rocks of Bohemia. Barrande supported this theory by reference to certain cases of recurrence of organisms after their temporary disappearance from an area. Such cases of recurrence will be discussed presently: as regards the actual theory, it has long been abandoned.

It is obvious that time must have elapsed before species originating in one area reached another remote from it, and the utility of fossils for purposes of correlation depends upon the length of time taken for migration as compared with the duration of the species. If the former was a large fraction of the latter, the organisms would be of little use; if, however, the time taken for migration was brief as compared with the duration, the method would be of great use. As in the case of human history we may usefully speak of events occurring at different hours of the same day as contemporaneous, so in the case of earth-history events occupying a few centuries may be regarded as contemporaneous when compared with events which occupied long geological periods.

Reasons will be given in a later section for believing that in the case of those fossils which have as
the result of experience been found useful for correlation over wide areas, the time of migration was
brief in comparison with that of the organisms,
while in cases where the dispersal took a long time
as compared with the duration, the organisms have
been rejected for the purpose now being considered.
The Zonal Method based on fossils is now regarded
as by far the most important means of dividing
and correlating strata, and most important stratigraphical work is now based on it. Much miscon-
ception has arisen with regard to this zonal method:
a zone simply means a belt, and the belts defined by
this method are determined by restriction of fossils
to definite horizons. Thus the strata are divided into
a number of fossil zones. The term has been largely
restricted to belts of small thickness, and thus has to
some extent aroused suspicion as to its utility. But
the belts are of very different degrees of importance,
and there is no reason why the thicker belts should
not be spoken of as zones as well as the thinner ones.
We might with perfect correctness speak of the whole
of the Palæozoic rocks as constituting the zone of
Trilobita, just as we speak of a few feet of Ordovician rock as the zone of *Nemagraptus gracilis*.
Thickness, then, is of no importance, and there is no
doubt that as regards duration of time our zones
have values differing greatly. In practice the term
fossil zone is applied to minor divisions, the major
ones receiving other designations.

Lateral Distribution of Organisms: Barriers

In the case of sediments, those which possess uniformity of characters over wide areas are most useful for purposes of correlation: so with organisms those forms which have the widest distribution in space are most useful for the same purpose. It is found that as with lithological characters, so with organisms, those of the rocks deposited farthest out from the coastal belt have similarity of character over the widest extent. This was unconsciously grasped in early days in the case of fossils, as witness the use made of the ammonites and the graptolites. Recently a much fuller study than was previously made has been carried out with reference to the distribution of existing forms of life, and it is necessary to pay some attention to this subject.

Biologists divide aquatic organisms into three main groups according to their conditions of existence, namely, the *Benthos*, comprising those organisms living on the sea-floor, which are further divided into the sessile and vagile benthos, the former comprising the forms attached to the floor of the water area, and the latter those which wander over the floor. Two other groups include the forms of life which live free in the waters above the floor, chiefly near the surface. These groups are spoken of as *Plankton* and *Nekton*, the former being the organisms which float involuntarily, the latter those which move more voluntarily, but the geologist generally includes both groups under the head of plankton.

In addition to the true plankton, other forms exist
under conditions which link them to the plankton
so far as distribution is concerned. The meroplankton
includes forms which, though benthonic in the adult
state, have free-moving larval states, while the
pseudoplankton contains organisms which, though
fixed, are attached to floating substances, usually
algae, and their dispersion is dependent on the factors
that control that of the seaweed. All these floating
forms are more readily spread than those living on the
ocean floor: accordingly the planktonic forms may be
expected to afford better aids to correlation than the
benthonic, and experience leads us to believe that this
is the case. The planktonic forms are carried far and
wide, often rapidly, by ocean currents, and the
physical conditions in the more central parts of the
oceans are often very uniform over wide areas, thus
minimising the effect of barriers, which will be discussed presently. Even the dead shells of such forms
can be carried for long distances, thus spreading the
relics of organisms beyond their range as living forms.
The pseudoplanktonic forms distributed as described
above, also have their dead remains carried beyond
their living range. Weed is often found flung upon the
shore with tubes of *Spirorbis* and many other organisms attached. The delicate limpet-like shells of
Helcion, either attached to the fronds or burrowing in
the stalk of *Laminaria*, are picked up on parts of the
coast, having been brought from their actual habitat
by the detachment of the weed. Masses of weed
accumulate to a great extent in the central swirls of

the oceans, as, for example, in the Sargasso Sea. The "gulf weed" extends over great stretches of the surface and, besides sheltering many true planktonic forms, has pseudoplanktonic organisms attached to it. Charles Lapworth gave valid reasons for believing that in Lower Palæozoic times some, at any rate, of the graptolites were pseudoplanktonic, attached to sea-weeds. This view was expressed by him in a contribution to a memoir by Prof. J. Walther, "Über die Lebensweise fossiler Meeresthiere" (*Zeits. deuts. geol. Ges.* 1897, p. 209), which may well be consulted in its entirety. Sometimes the pseudoplanktonic form is not attached to the float, but the float to it. An instance occurred a few years ago in the Scilly Isles, as recorded by Mr C. J. King, where a living crab, 3 in. across, carried five distinct branches of seaweed, the longest of which was 6 ft. 4 in. in length.

Though planktonic organisms are most useful for purposes of correlation, the benthos is by no means negligible, and their value is on the whole greatest when deposited far away from the coast-line, where conditions are often uniform over wide areas, allowing of free distribution on the ocean floor, without arrest by barriers.

Before discussing these barriers, however, it is important to consider the rate of distribution of organisms, for, as before stated, their value for correlation of strata depends on the brevity of the period of their distribution as compared with the duration of the forms. That freely floating forms can be rapidly transported for long distances is indicated by the occasional transport of seeds across the Atlantic by ocean currents in so brief a period that they have not lost their powers of germination. Similarly any planktonic form is liable to be drifted rapidly and for long distances. But benthonic forms may also be spread rapidly, if conditions are favourable. A species of *Littorina* was introduced into Canada from Europe in the last century, and in the course of a few years was able to spread along the coast for some hundreds of miles southward. The significance of this example is that, owing to the cold current along the coast, temperature variations were not great over several degrees of latitude, and the spreading creature found a similar environment far south of its starting-point.

A few years ago some very significant experiments were made at Beadnell on the Northumbrian coast. A number of marked crabs were liberated, many of which were subsequently captured on the coast at different times and places. One had travelled 155 miles in just under two years, and one had crawled at the almost incredible average speed of a mile a day! At this rate the progeny of such a crab might spread round the world in the course of a human lifetime, if there were no barriers, showing the very brief time required for dispersal, as compared with the duration of the species, granted favourable conditions. Conditions, however, are often unfavourable, owing to the existence of barriers stopping the further spread, temporarily or permanently. If temporarily, the spread will be slow, and may take a very long time, rendering the species of no use for purposes of correlation. If permanently, the ultimate areal distribution of the species will be limited, and it will be useless for correlation, save in that limited area.

Barriers. These, as is well known, are of many kinds, some physical, some physiological. Among the more important factors affecting aquatic organisms are temperature, light, salinity, purity of water, supply of food, hostile forms of life and overcrowding.

Barriers as a whole are of most frequent occurrence in the shallows along the coastal belts. Temperature variations are rapid, both laterally and vertically, light decreases downwards from the surface, salinity is different in estuaries, in partly enclosed seas and in the open ocean, and in the ocean itself the salinity varies under different climatic conditions. The water is in places clear, elsewhere rendered turbid by sediments of different kinds, while certain impurities may actually act as poisons. As variations in the constitution of the sediments themselves act as barriers, and these variations are specially numerous in the shallows, the organisms here show specially restricted habitats. Thus we find along the British coast that the limpet is confined to the rocks, the mussel to a habitat on gravel, the cockle and razorshell occur in sand and *Pullastra* in mud. Variations in the food-supply and other conditions are also frequent: all these things tend to check the spread of organisms in the shallows, causing them to have a local distribution.

In the parts of the sea-floor away from the land conditions are more uniform. Below a certain depth temperature tends to become uniform through many degrees of latitude. Salinity is usually fairly constant: at no great depth the effects of solar light cease: sediments are also of uniform character over wide tracts of the sea-floor.

Owing to this uniformity of conditions over large tracts of sea-floor, the benthonic forms are less checked by barriers than those of the shallows, and accordingly we find that they are of very considerable use in enabling us to correlate sediments deposited away from the coast-line over wide areas. It is in fact found that, in studying the sediments of past times, it is difficult to make direct correlations of the variable sands, silts and shallow-water limestones deposited in a coastal belt, but easier to do the same in the case of the fine muds and open-water limestones formed far from the ancient coasts.

With planktonic forms conditions of existence are yet more uniform over wide areas; the upper waters are practically free from pollution by sediment away from the coast-line, and the only varying factor of importance is temperature. This, however, is probably more than counterbalanced by the fact that conditions favour rapid and wide dispersal of both living and dead forms. It is to the plankton that we have to appeal in order to obtain the greatest assistance in correlating strata by their included organisms.

Establishment of Zones: Zone-fossils

The various zones have been established as the result of observation. They are of very different values as regards thickness and horizontal extent. The selection of suitable zones is the result of experience. The best for most purposes are those which have a wide horizontal extent, and are of relatively small thickness, having been accumulated in brief periods of time. Some of these are practically world-wide, as is the case with certain graptolite zones of the Ordovician system. Even when zones are defined by the use of organisms of the same group, the various species characterising the zone may have very different ranges in time. For example, the Valentian zone of *Monograptus gregarius*, originally defined by C. Lapworth, has been subdivided into minor zones, each characterised by other species of *Monograptus*, which have shorter time-ranges than *M. gregarius*. The ranges in space also differ in a marked degree. For example, the zone of *Nemagraptus gracilis* is found in Europe, Australia and America, while that of *Monograptus argenteus* has only been found in a small part of Britain.

Each zone is usually characterised by one particular fossil, termed the zone-fossil. It must not be supposed, however, that the name-fossil of the zone is the only organism characterising it. It is often selected in an arbitrary way, and other forms might equally well have been chosen. In fact a fossil zone is actually identifiable by its assemblage of fossils, of which the name-fossil is but one. In distant areas it would be dangerous to make correlations on the strength of one organism only. When satisfactory correlations are made, they are the result of the establishment of sequences of zones, each containing several organisms common to different areas of distribution and not found in the zones above and below.

The name-fossil of the zone will naturally be abundant in the place where the zone was defined and named after it, but in other places it may be very rare or even absent. In the latter case other fossils must be used for purposes of correlation. The instances where it is rare are often of high interest, as we may be able to correlate strata deposited under very different conditions. At the present time a planktonic univalve mollusc, *Ianthina*, lives in the surface waters of the open Atlantic in great numbers, and its shells must be accumulating in great quantity on the floor of the ocean, where it will become a characteristic fossil for future use. The shell is not found living in the British seas, but dead shells are occasionally washed on to the shores of Devon and Cornwall, and in larger quantity on those of the Channel Islands. Accordingly, a future geologist who studied the deposits now being laid down in the open Atlantic, and the British shores, would find very different assemblages of organisms in the two areas, but the *Ianthina* shells would give him the clue as to their contemporaneity: the two sets of deposits could rightly be spoken of as the zone of *Ianthina*.

A similar distribution in ancient rocks actually

enabled C. Lapworth to correlate the later Ordovician
and earlier Silurian rocks of the Moffat and Girvan
tracts in the Southern Uplands of Scotland, though
deposits are very different and their faunas essentially dissimilar. The deposits of Girvan are of
shallow-water origin, with a fauna of what is termed
the shelly facies, while the Moffat sediments are of
more open-water character, containing little but
graptolites. The graptolitic deposits were divided by
Lapworth into a number of zones, and a sufficient
number of graptolites have been found in the Girvan
area to enable the deposits there to be correlated zone
for zone with those of Moffat, although the faunas are
otherwise entirely different.

This case enables us to illustrate another matter
connected with fossil zones. Those who originally
defined the zones usually ascertained the sequence
where the strata were exceptionally thin. In such
places it was often easy to ascertain a sequence of
half a dozen zones in a single cliff-section, whereas
where the strata accumulated in the same period were
of much greater thickness, a considerable amount of
ground might have to be traversed before the same
number of zones were established. Accordingly an
idea arose that these zones were usually found in
great numbers in a small thickness of deposit, and
this increased the suspicion held in some quarters as to
their validity for purposes of correlation. Actually,
however, the zones, when traced laterally, are often
found to expand into much greater thicknesses. The
strata of the Scottish Southern Uplands are an

example. The shallow-water deposits of the Girvan area, which are equivalent to the graptolitic shales of Moffat, have a thickness of over 3,500 ft., while the Moffat shales are under 300 ft., so that the thickness diminishes at Moffat to less than a tenth of that found at Girvan. This is only one of many instances.

Fortunately, by the use of planktonic (including the pseudoplanktonic) forms, the greater part of the fossiliferous rocks can be divided into zones for purposes of correlation, for when the graptolites disappear towards the end of Lower Palæozoic times, the work can be carried on through Upper Palæozoic and Mesozoic rocks by means of the ammonoid cephalopods, generally regarded as planktonic. The Cainozoic rocks present greater difficulty, owing to the comparative rarity of planktonic forms in what are to a large extent coastal deposits in the areas where they have been most fully studied.

After the planktonic forms the next in importance are the benthonic fossils of the open-water tracts, away from the coasts. The trilobites of these deposits are of much value throughout Palæozoic times. When they disappear evidence is afforded by other organisms, of which the Echinoidea are specially useful. Even in the case of the benthos of the shallows, a certain amount of evidence may be afforded, particularly in the case of the meroplanktonic forms having floating larval states.

As our knowledge progresses we shall be able to utilise the various groups yet more successfully, for we have still much to learn about the species or varieties of many groups. Fossils which have been referred to one species having a long time-range are often found to be divisible into a number of species or varieties, each having a more restricted duration. This is for instance the case with the brachiopods, which are now found to be of considerable value for purposes of correlation.

No doubt, when further perfecting the study of correlation by zones, the question of lines of evolution will have to be taken into account. This has already been attempted to some degree by several writers. It will probably be of much use in the future, but is still fraught with danger on account of the difficulties associated with attempts to establish phylogenetic branches.

Further Objections to Zonal Correlation: Recurrences

The objections to detailed correlation by fossils depend, as we have seen, on the rate of dispersion from specific centres as compared with the periods of existence of the organisms. If the time for migration is relatively long the sequence of zones is only indicative of homotaxis; if short, of practical contemporaneity.

That all kinds of organisms are not of equal value for purposes of correlation soon became apparent. An example may be given where correlation cannot be made. Two genera of trilobites of the *Phacops* group appear in early Ordovician rocks in Europe, namely, *Chasmops* and *Dalmanites*. *Chasmops* is first found in Esthonia and Scandinavia, and does not appear in

Britain until Upper Ordovician times. It has not been found south of Belgium. *Dalmanites* appears in Spain, France and Bohemia in early Ordovician times, and only reaches Britain and Scandinavia at the end of the period. Accordingly *Dalmanites* appears in strata newer than those containing *Chasmops* in Britain, while if *Chasmops* had migrated further southward than it seems to have done it would be found in strata overlying the *Dalmanites* beds of Spain. It has been pointed out that such slow and probably interrupted dispersal is prone to occur under the conditions in which the coastal deposits were laid down. Experience soon teaches us to avoid such organisms when attempting detailed correlation. The fact that other organisms do not show any such indication of slow dispersal at once suggests their greater value as aids to correlation. The ascertainment of the age of the *Chasmops*- and *Dalmanites*-bearing beds in various regions depends on the possibility of fixing the ages by the faunas as a whole, which enables us unhesitatingly to assign the various beds to their proper horizons.

We may now briefly consider the difficulties which confront us if we regard the zonal sequences as not indicating contemporaneity, but merely homotaxis. Even if we assume that successive faunas arise in one place only and migrate in one direction at equal rates, a fauna starting from that place and travelling round the world should appear there in strata overlying those containing faunas of later origin. Recurrences certainly are found and will be considered presently, but they can be accounted for in a different manner.

In actuality faunas do not originate at the same place, but arise in various areas, spreading from thence not along single lines but radially outwards from the point of origin. This being so, the succession of faunas in time will become hopelessly confused when compared in various regions. That instances do occur has been already shown in the case of the Phacopidae, but these are shallow-water benthonic forms, which are not generally used for purposes of correlation, whereas in the forms so used the evidence points most strongly to the fact that the non-recurrence of zonal assemblages of different horizons is due to rapid dispersion compared with the duration.

As has already been stated much has been made of recurrence by opponents of detailed correlation. These recurrences are very common. They are well known as occurring on a small scale in the course of a human lifetime. Many species of butterflies visit certain countries in some years in vast swarms, being apparently absent in the intervening years. Similarly we find fossils occurring in certain strata apparently absent from the succeeding ones and reappearing in still later beds. The apparent absence may be real. The forms may disappear from the particular area in the interval, or they may linger in such diminished numbers as to escape detection. The disappearance is often clearly shown to be due to some physical change. Good examples of such recurrences are found in the Palæozoic rocks of Bohemia, and were long ago recorded and discussed by Barrande. The Ordovician rocks were divided by him into five "bandes", indicated by the letters *Dd* followed by the figures 1–5, in upward succession. Of these, *d* 1, *d* 3 and *d* 5 are open-water sediments, while *d* 2 and *d* 4 are of shallow-water origin. Accordingly certain forms are found in *d* 1, *d* 3 and *d* 5 which are absent from *d* 2 and *d* 4, and vice versa. The trilobite *Aeglina rediviva* is found in *d* 1, *d* 3 and *d* 5. This is a case of a form having an unusually long time-range, with temporary local disappearances. This is no real difficulty, for the cases of recurrence are few, as compared with those of fossils having a continuous time-distribution. Further, there is no real anomaly: the case of *Aeglina* is strictly comparable with that of *Monograptus gregarius* already noted, in which the zone is split up into minor zones, so *Aeglina* may be taken as characteristic of the Ordovician zone as a whole, while the minor subdivisions of the Ordovician have their own zonal fossils, e.g. *Phillipsinella parabola* is the zone-fossil of *d* 5.

One interesting example of recurrence may be given by reference to a wider area. The trilobite *Ampyx* is abundant and widespread in Ordovician times, but is found but rarely in Silurian rocks. It occurs in the Valentian of Britain and Scandinavia and in the Lower Ludlow rocks of Britain and Bohemia, but has never been found in the Wenlock, though it doubtless will be found in this series eventually. We may here be dealing with an actual case of world-traverse, followed by reappearance in its old region, though this is by no means necessary and is against the known facts of its distribution.

One test of the reliability of the zonal method of

correlation has been frequently applied. Collections of fossils from distant countries, supposed to belong to one horizon, have been submitted to experts, who have been able to establish the existence of several zones by identifying the fossils on separate slabs. It has been found that where a zone-fossil is associated with other forms on a slab, these all belong to that zone, while on other slabs each zone-fossil is so accompanied by its proper associates. These determinations have sometimes been disputed by those who obtained the fossils, only to have the correctness of the determination subsequently admitted. I have myself made out the graptolite zones of the Fichtelgebirge in this way and compared them with those of Britain. Similar determinations from regions so remote as Canada, South America, Australia and New Zealand have been made by others, notably by C. Lapworth, Miss G. L. Elles and Dame Ethel Shakespear.

A warning must be given against the possibility of errors due to confusing *remanié* fossils with those actually contemporaneous with the deposit: such confusion has been made in the past, and may be made again in the future unless sufficient care be exercised. As an example of error arising in this way the Cambridge Greensand may be cited: it is really Cenomanian, but was once regarded as Albian on account of the preponderance of fossils of that age over those indigenous to the deposit.

The reliability of the zonal method is now fully admitted by geologists, it being understood that the necessary discretion is exercised in its use.

CHAPTER IV

DEPOSITS ON LAND AND SEA

HAVING considered the use of lithology and fossils for purposes of correlation, we pass on to discuss their utility as indices of prevalent physical conditions at the time of the formation of the deposits. Firstly, attention may be paid to differences of lithology and of organic contents on land and sea respectively. The deposits with which the geologist has to deal are mainly marine, but those of land-formation are of much interest, not lessened by their comparative rarity over many regions. They may be strictly terrestrial, or fluviatile and lacustrine, the last being accumulated in either freshwater or salt lakes.

Beginning with lithological characters, it may be noted that although there are criteria available for distinguishing terrestrial and aquatic deposits, it is difficult on lithological grounds alone to discriminate between freshwater and marine deposits. Speaking generally the coarse pebble-bearing deposits are marked by greater rounding of marine pebbles than those of freshwater origin, and of the latter than the truly terrestrial accumulations. The pebbly accumulations of the dry land are of the nature of rubbly material, the large fragments, not being worn by water action, remaining angular. Such are the screes and detrital accumulations of the high ground and their fringes. They are often marked by much ferruginous material in the matrix. It must be noted that rounded pebbles may be incorporated in such deposits, having been derived from pre-existing pebble beds: they are in fact comparable to *remanié* fossils. Stratification is generally absent, and when present irregular: indeed we may say that the deposits formed on the land are generally distinguishable from those deposited under water by absence or irregularity of bedding in the former and presence of the same in the latter, though there are many exceptions, as the well-bedded sands of the desert dunes. The truly terrestrial deposits often contain a considerable amount of peaty and ferruginous matter, though this again is not a general criterion, for it may be absent, and on the other hand both may be abundant in some aquatic deposits. In fact the so-called "Forest Bed" at Cromer, with abundant peaty matter, was once supposed to be an old surface soil, but is now known to be marine. The long controversy that has raged concerning the terrestrial or aquatic origin of coal, with its large percentage of vegetable matter, illustrates the difficulty of accounting for such accumulations.

In distinguishing fluviatile from terrestrial accumulations the following features may be noticed. They are often marked by a linear arrangement across country; the pebbles will probably be water-worn to some extent, and the fine argillaceous material will be stratified and even laminated. None of these tests may be available in any given case, and we are ultimately driven to the organic contents for definite confirmation of origin.

In freshwater lakes the pebbles of the beaches vary in degree of wear: they are often comparatively little rounded, but this may also be the case with beaches formed in the tranquil embayments of the sea, and especially in fjords and rias. Most of the deposits are in character very like those of fluviatile origin, though the finer sediments bulk much more largely than in the case of river deposits.

The deposits of salt lakes are more characteristic: chemical precipitates, as gypsum, rock-salt and more rarely the highly deliquescent chlorides and sulphates of potassium and magnesium occur here. Owing to evaporation also tracts of mud and silt are laid bare and are marked by sun-cracks, rain-prints and tracks of crawling animals. Pseudomorphs of mud after cubes and "hopper-crystals" of rock-salt are characteristic.

The marine deposits are of mechanical or organic origin, with rare exceptions to be considered later. Further details of their lithological characters will be considered in a later chapter.

It will be convenient here to notice the occurrence of false-bedding. It is mainly though not exclusively of aquatic origin; it also occurs in the wind-drifted sands of the terrestrial dunes. When aquatic it is exclusively of shallow-water origin, not being found in the great depths of the oceans. It is indicative therefore of terrestrial dunes and of fluviatile, lacustrine and shallow-water marine conditions, and is important inasmuch as when we meet with it we know that we are not confronted with deep-water sediments.

It will be seen that lithology can on the whole only be used as subsidiary to the information derived from an examination of the various organisms contained in the deposits. To the consideration of these organisms we must now turn. At the outset it must be noted that dead remains of organisms may be carried from their proper habitat to another type of physical surroundings. Thus land animals and plants may be carried into rivers, lakes and the sea, and river inhabitants into the sea: they are usually transferred from a higher to a lower tract, but not invariably, for sea-fowl often carry marine shells on to the land. We must expect therefore to meet with organisms in deposits formed under conditions other than those of their proper environment. Sometimes such relics are numerous. In certain river-gravels land-plants, land mollusca and mammalian bones are more frequent than the fluviatile organisms proper to the deposits: indeed the latter may occasionally be completely absent.

The case of kitchen-midden mounds, consisting largely of the remains of edible mollusca piled up by men, is a case in point, and one worthy of mention, for these heaps have sometimes been mistaken for raised beaches.

With this warning, we may now pass on to consider the organisms characterising the deposits formed under the varying conditions which we are now discussing, taking first those found in the accumulations on the dry land. The fossils of the land accumulations will, with very rare exceptions due to such causes as those noted above, be land-dwellers. Roots, branches, leaves and seeds of plants are often abundant. Of invertebrate animals, mollusca, galley-worms and insects are not uncommon. The mollusca are entirely univalve, pulmoniferous gastropods, with entire apertures. Mammalian remains are not infrequently found among the vertebrates. Any of these organisms may be and frequently are carried into rivers and lakes, where they are often very abundant, and accordingly the positive evidence supplied by the presence of the above-noted relics in accumulations is not convincing proof of their terrestrial origin. Negative evidence furnished by the absence of aquatic organisms is useful, but in itself is not sufficient proof of terrestrial origin. The most satisfactory proof of such origin is supplied by the existence of roots of land-plants ramifying through the material, and still retaining their ultimate rootlets, thus indicating that the plants grew *in situ*. This point is of great importance, for tree stumps may be washed from their native soil, float down rivers and, becoming water-logged, sink to the bottom in a river, lake or the sea, giving the deceptive appearance of growth *in situ*. This occurs in the aforementioned Cromer "Forest Bed." Here the ultimate rootlets are absent, and only the broken major roots remain, rounded at the broken ends by water wear. As many aquatic plants have rootlets ramifying through the mud at the bottom of a water tract, we must also make sure that the plants with which we are dealing are actual land-plants. This is usually easy in the case of fairly modern deposits, but in the case of ancient strata in which the included plants are extinct forms, it is a more difficult matter. The controversy concerning the origin of coal, for instance, was complicated by the doubt as to whether the plants occurring *in situ*, with rootlets penetrating the underclays, were of terrestrial or aquatic habitat. In beds of later age we can often ascertain the terrestrial origin with certainty, knowing that the plants whose roots are *in situ* are of terrestrial growth. Certain submerged forests of early Holocene age found along many parts of our coasts can, unlike the Cromer "Forest Bed," be proved to be of terrestrial origin by the frequent abundance of rhizomes of *Osmunda*, penetrating through the forest bed, which is an actual surface soil.

Turning now to the freshwater deposits, we find that numerous terrestrial organisms may be embedded in them, but in addition truly aquatic forms are almost invariably associated with them, and these are usually more abundant than the land forms. In the case of comparatively modern strata the presence of living species or genera of freshwater habitat will indicate their origin. For example, the freshwater beds of the Tertiary strata frequently contain such genera as *Vivipara* and *Planorbis*, and we also find such living genera even in the Cretaceous and Jurassic. In the case of deposits of a more remote period, however, we are often left in a state of uncertainty. Apart from identification of genera a certain amount of information may be supplied by the mollusca. The terrestrial mollusca are all univalves, whereas in addition to these bivalve lamellibranch shells live in the freshwater areas. The presence of these in a deposit will indicate that it is not of land-formation, though it may be marine.

This brings us to the distinction between freshwater and marine deposits. Taking first the lamellibranchs: on the interior of either valve are certain impressions known as muscle-scars, being the parts of the shell to which the adductor muscles used in closing the valves are attached. There may be only one in each valve, as in the case of the oyster, or two as in the freshwater mussel, the sea-mussel and a host of others. The shells with only one scar in each valve are known as monomyary, and those with two scars as dimyary. Monomyary mollusca live only in marine or estuarine, and not in freshwater areas, therefore the occurrence of monomyary shells in a deposit indicates that unless the shells were introduced adventitiously, the deposit is not of freshwater origin. But their absence does not prove the freshwater nature of the deposit, for the larger number of marine shells are likewise dimyary, and it is quite usual to find marine deposits which furnish no monomyary shells. Assistance is also afforded by the gastropods. Some of these have an entire aperture, while in others the aperture is notched, frequently prolonged into a canal. The former are known as holostomatous, the latter as siphonostomatous. The holostomatous shells are those of terrestrial or freshwater, but very rarely of marine habitat, while the siphonostomatous shells are almost exclusively marine.

By aid of the lamellibranchs and gastropods, therefore, strong but not always conclusive evidence is supplied as to the freshwater or marine character of the containing deposit.

Considering the faunas as a whole, the freshwater areas are marked by the absence of a large number of the major groups of animal life which are found in the seas.

The following groups are probably confined to the seas, or at any rate such of them as secrete hard parts. The sponges may be regarded as indications of marine deposit, as the freshwater sponges have no hard parts. Important forms which are exclusively marine are the corals, sea-urchins, crinoids, starfish and brittle-stars, barnacles, brachiopods and polyzoa, and among the mollusca the pteropods and cephalopods.

It is not often that we meet with a marine deposit without some of these, so that the determination of a marine deposit is usually a matter of ease. When however none of them is met with, it is uncertain whether the deposit is marine or freshwater, and although such cases are rare they do occur. Occasionally deposits are found containing dimyary lamellibranchs to the exclusion of all other fossils, and it is then a matter of doubt as to whether the containing deposit is fresh water or marine. As actual cases, two examples may be cited. In some beds of the Old Red Sandstone of Britain, a lamellibranch, *Archanodon jukesii*, occurs alone, and in certain beds of our Coal-measures other lamellibranch shells of the genus *Naiadites* and allied genera are also found unaccompanied by other fossils. *Archanodon* and *Naiadites* are dimyary, and it is a matter of dispute in the case of both Old Red and Coal-measure deposits containing them whether they are of freshwater or marine origin.

From what has been written above it will be inferred that by making a careful study of the included organisms, and supplementing this information by evidence derived from examination of lithological characters, it is rare to be left in doubt as to whether any particular deposit is of terrestrial, freshwater or marine origin.

CHAPTER V

ACCUMULATIONS AND DEPOSITS OF THE LAND AREAS

Areal Distribution

It is not my purpose to give a very detailed account of the lithology and organic contents of the accumulations and deposits formed upon the land tracts. They are of very diverse character, and a full description would require more space than can be devoted to them here. Even in a small country like Great Britain, of which the different areas are marked by general similarity of climatic conditions, such diverse accumulations are formed as the soils and peats of the uplands, the alluvia of river valleys, and the dunes of the coast, among a host of others, and it is obviously impossible to describe all the variations found throughout the world. Many books give detailed accounts of the lithology of the terrestrial deposits, and for those who require further details special reference may be made to G. P. Merrill's *Rocks, Rock-weathering and Soils*. The most satisfactory method of treatment on broad lines is based on climatic conditions as controlling factors in the formation of the deposits. The great climatic belts into which the world is divided leave their impress on both the lithology and the organic contents of the deposits formed in them, and useful lessons are thus afforded as to climatic conditions in past times, fullest in the case of the deposits of late geological date, but by no means negligible in the case of earlier formations.

Some of the features so far as the belts furthest removed from the equator are concerned were many years ago treated by A. Nehring in a paper published in the *Geological Magazine* for 1883, p. 51.

It is true that conditions due to differences of latitude are imitated on a small scale in the various belts of altitude, passing in equatorial regions from tropical to glacial conditions, but on a large scale they are distinctive of climatic belts. These belts, running roughly parallel with the equator, are more accurately defined by isothermal lines than by latitude. In each belt there is great variation in detail as regards character of the deposits and of their organic contents, but each has certain features which are essentially characteristic, though not necessarily entirely confined to it, since they may occur elsewhere to a less marked degree. Further, these features need not necessarily be found over the whole area, but nevertheless they form a dominant note in the belt.

The belts, according to these dominant notes, may be designated as follows, proceeding from the poles to the equator:

1. Glacial.
2. Tundra.
3. Steppe.
4. Temperate forest.
5. Subtropical desert.
6. Tropical forest.

Briefly the main climatic conditions in the various zones are as follows:

Glacial, cold and dry in its most typical development, the dryness being due to precipitation of vapour as snow and not as rain.

Tundra, cold and damp, the dampness being due to melting of snow and ice.

Steppe, fairly cold and dry, the dryness due to interior continental conditions.

Temperate forest, fairly warm and moist.

Subtropical desert, very hot and dry.

Tropical forest, very hot and damp.

Each belt does not necessarily form a continuous zone across the lands; any one may be interrupted for some distance, but they occur always in the order above enumerated. The same is true as regards sequence in time. The belts are more complete in the northern than in the southern hemisphere, owing to the deficiency of land in the southern part of the latter, and as the deposits of the northern hemisphere have been more fully studied, examples will be mainly though not exclusively taken from it.

The belts will now be considered in detail, starting at the poles. The lithology and nature of the organic contents of each will now be considered. The terms used in Merrill's classification of the regolith will be adopted, the accumulations being distinguished as Sedentary and Transported, the former including the subdivisions known as Residual and Cumulose, and the latter Colluvial, Alluvial, Glacial and Aeolian.

The deposits of the lakes will also be treated as alluvial, since the term is used in Europe to include lake deposits as well as river deposits, though in the United States it is frequently confined to the latter.

Description of the Belts

(1) *Glacial Belt.* Proceeding now with a detailed consideration a beginning will be made with the glacial belt. The glacial belts are situated around the two poles. That around the South Pole is more typical owing to the Antarctic continent, which has allowed of the development of the great circumpolar ice-sheet, occupying the greater part of the area. Around the North Pole is much sea, and the masses of land ice are less extensive, occupying parts of various islands, of which Greenland is far the most important, and possesses an ice-sheet resembling in a general way that of Antarctica, save in size.

The typical accumulations of this belt are the products of the ice-sheets and glaciers, especially boulder-clay and moraine-material. In ice-free areas other materials are accumulated, bearing the impress of glacial conditions. The rivers discharged from the ice carry a load of material of differing degrees of size, varying from large ice-transported boulders to very fine mud produced by subglacial erosion. These materials are accumulated along the river courses and in lakes as aqueo-glacial deposits. The boulders often retain the characters impressed upon them by the ice, namely, their subangular outline and the striations due to ice action. A frequent feature is the inclusion of lenticles of fine glacial mud with the coarse detritus, the mud being laid down in pools and the still waters of cut-off braided channels in the flood plains often found in advance of the ice-margin; much of this mud is also deposited in lakes. The mud is usually well laminated, often exhibiting the striping of "varve clay" when formed under certain conditions, as described by Baron de Geer[1]. It is also usually very tenacious owing to the minuteness of the component particles, and is frequently therefore referred to as gutta-percha clay.

In addition to the material deposited directly or indirectly from the ice, other deposits are found, but they are not characteristic. Of these, residual deposits are unimportant, for the general absence of vegetation allows agents of transport to remove them as quickly as they are formed by weathering; therefore few are preserved. The general absence of vegetation also forbids the formation of cumulose deposits such as peat, save under special circumstances. Colluvial material is formed in some abundance in the ice-free tracts, especially screes and hill-detritus of a finer character. In the ice-free land tracts conditions often occur on a small scale resembling those dominant in the tundra belts, and their characteristics will be dealt with when considering those belts. One feature in

[1] Geer, G. de, "A Thermographical Record of the Late Quaternary Climate" in *Die Veränderungen des Klimas*, Geol. Congress, Stockholm, 1910; "A Geochronology of the last 12,000 Years," *ibid*.

the aqueous deposits may be noticed, which occurs
to some extent in the fluvio-glacial deposits, but is
more marked in those of the lacustro-glacial origin. In
normal aqueous sediments the materials vary in size
with the velocity of the transporting currents, and
accordingly in any particular layer we do not meet with
any great departure from uniformity. The waters of
cold regions, whether of rivers or lakes, often bear
masses of floating ice which can transport fragments
of much larger size than those which are being transported by the water at the same time. Hence we may
and often do meet with large boulders embedded in
muds of the very finest grain. Such an occurrence is
not confined to the glacial belts: it sometimes happens
even in temperate climates, such as Britain, in exceptionally severe winters, but only on a small scale,
whereas in the glacial belts it is developed in a very
high degree.

In this account of the accumulations of the glacial
belt it has been assumed that the reader is acquainted
with the characters of the normal glacial deposits, as
boulder-clay and moraines: they have therefore not
been described at length, whereas a good deal of space
has been devoted to the subsidiary deposits, whose
characters are not so generally known.

In the glacial belt contortion is produced in more
than one way. The movement of ice-sheets and glaciers
causes it in deposits underlying them, and even in
those frozen into the lower layers of the ice itself. In
aquatic deposits it may be brought about by the
grounding of masses of floating ice upon them, and

in another way it may be due to thawing of layers of ice intercalated between deposits.

One important feature, characteristic of the deposits of the glacial belt as a whole, remains to be considered. Owing to the scanty rainfall, weathering is mainly due to physical rather than chemical agencies, and attrition by ice is also purely mechanical: therefore the deposits of this belt as a whole are marked by the retention of most of their soluble material. In this they resemble those of arid regions.

Little life exists in the portions of the glacial belt actually occupied by the ice, though there is abundance of vegetable and animal life in the ice-free parts. As this is essentially similar to that of the tundra belts, it will be considered under that heading.

(2) *Tundra Belt.* The transition from the glacial belt to the tundra belt is not so regular as it would be if it depended on latitude alone. Altitude also plays an important part. It has been seen that some low-lying tracts of the glacial belts are under tundra conditions and conversely uplands of the tundra belt may be glaciated. The line of demarcation therefore is complex. In the southern hemisphere the tundra belt is inconspicuous as the glacial belt is mainly bounded by sea to the north, and this is so in a less degree in the northern belt, for its principal development is in northern Canada and northern Siberia, each of which is bounded to the north by the Arctic Ocean. In both the old and the new worlds the existing tundra belt is largely of the nature of a coastal plain, having emerged from the sea at a late period, and is therefore mainly

low-lying. The great rivers in Siberia and in Canada flow northward and the waters near the sources are thawed in spring while those of the mouths are still frozen. The waters towards the south, swollen by the products of the melting snows and blocked by the frozen waters nearer the mouths, overflow their banks and flood the surrounding plain, giving rise to the swampy tracts which are one of the main character- istics of the tundra region. Another circumstance affects the character of the tundra deposits of the present day. During the Pleistocene period the glacial belt extended far south of its present termination, and accordingly glacial muds were deposited as boulder- clay and other accumulations in what is now the tundra region, therefore the present tundra deposits contain a larger proportion of fine material than would otherwise be the case, this being due to the erosion and redeposition of these glacial deposits.

It is to the coldness in addition to the wetness of tundras that their characteristic features are due. Though the surfaces are often marked by morass and quagmire in the summer, the typical tundras are per- manently frozen at no great depth below the surface, and the surface is itself frozen during a considerable part of the year. The frozen layers below the surface prevent downward drainage and so increase the super- ficial moisture of spring and summer. Surface water frozen over may be in that state covered with fresh sediment during spring floods, and by degrees alter- nations of ice and sediment are so produced. If a lakelet be thus buried, a mass of ice of considerable thickness and horizontal extent would utimately be embedded with the sediments. Subsequent melting of embedded ice would give rise to considerable disturbance and even contortion of the deposits. Masses of frozen sediment and peat would become detached in this and other ways and might be transported for some distance, and ultimately incorporated as boulders among later-formed sediments. Animals would be trapped in the morasses and their carcases ultimately frozen in and preserved entire. The Abbé Huc describes a case of a number of yaks having actually been frozen suddenly into the ice while swimming a river in Central Asia, their frozen heads only appearing above the surface. Similar occurrences must take place in the rivers of the tundras and the ice imprisoning them may be covered up by subsequently formed strata.

Having thus described the physical characters of the belt, we may consider its deposits in further detail. The sediments of the flood-plains and river beds have naturally the general characteristics of those of other regions, namely, deposits of varying degrees of coarseness in river beds, and finer material on the floodplains, and these need not be further noticed; we may confine our attention to the features special to the region. The wide extent of the plains permits of an exceptional amount of fine sediment being accumulated as laminated mud and silt, resembling in some respects the muds of the glacial belt, but on the whole less fine. These will be intercalated with lenticular masses of gravel and sand, owing to lateral shifting of the river beds. A very important feature is the frequent intercalation of thin peaty seams, due to abundant growth of vegetation in the wet seasons. The peat itself differs in a marked degree from that of temperate regions, being composed largely of lichens and mosses; it is very fibrous and usually of a light colour, strongly resembling some forms of cut tobacco. Terrestrial and freshwater shells are found in mud and peat alike. The distribution of boulders by floating ice, in the manner previously described, is significant; especially important as indicative of cold conditions is the transport of boulders of frozen material, which would otherwise be incoherent. Contortion may be produced in the sediments after their deposition; one cause has already been mentioned; another is due to grounding of floating ice-blocks during the thaw.

As in the case of other belts special modifications occur on the high grounds margining the flats, where colluvial material is found, including rubble-drifts. In the tundra area these may, like the deposits of the flatter tracts, contain blocks of material transported in a frozen condition, becoming incoherent again when thawed.

Turning now to the organisms, there are certain features of importance connected with the vegetation, especially as regards the leaves. Both here and in the glacial belt they are modified to resist the periods of dryness alternating with the moisture of the flood times. The leaves are protected against this dryness in various ways, so as to present as little surface as possible to transpiration: they are frequently needle-like or strap-like, and may be curled over into hollow cylinders, to protect the stomata beneath, which may be further protected by hairs. The substance of the leaf is often leathery and some possess a waxy surface. There are other modifications which are not so readily detected in the case of fossil leaves. The trees are usually dwarf and the leaves are therefore small as compared with those of the same genera in warmer regions, as well displayed by the Arctic willows and birches. It should be noted that some of the modifications displayed by leaves in the tundra are also found in plants growing in parts of the subtropical deserts, which also have to endure periods of drought, but these are not likely to be found in the fossil state.

The modifications of animals are not so pronounced, though many of the mammalia are provided with specially thick hairy or woolly coverings.

As it is an important task of the geologist to restore the physical conditions which prevailed in various countries in Pleistocene times, when the organisms were to a large degree specifically identical with those now living, brief mention may be made of some of the more important existing tundra forms. Many of these are found, for example, in Pleistocene beds in Great Britain. Numerous species of what is sometimes spoken of as the Alpine, sometimes as the Arctic flora occur. Those of special interest are *Dryas octopetala*, the dwarf willow, *Salix herbacea*, and the dwarf birch, *Betula nana*. Both land and freshwater mollusca occur; special notice may be taken of the highly Arctic terrestrial *Columella columella*. Of the mammals the reindeer, musk-ox and marmot may be noted.

(3) *The Steppe Belt.* The two belts described above conform much more closely with the latitude than the two that follow. The dry conditions necessary for the formation of steppe accumulations are produced not so much by a particular latitude as by the existence of continental climatic conditions of aridity, found in the interior parts of the great land masses, while on the other hand the conditions of humidity necessary for the vegetation of the temperate forest belt are found in areas with an "insular" climate in the marginal parts of the same land masses. Hence these belts depart to a considerable degree from the lines of latitude. The steppe belt is found in central Asia and southern Russia, and is absent from the more western parts of Europe, while the forest belt, though well developed in western Europe, becomes less clearly defined further to the east.

The steppe belt is of particular interest to geologists on account of its much greater areal extension during parts of the Pleistocene period, as shown by Baron F. Richthofen in his well-known treatise on the Loess[1]. This was formed over very wide areas in Pleistocene times when those areas were under steppe conditions, and the account given of the steppe deposits in this chapter is largely taken from his paper, while the fauna is discussed by Nehring in the work already cited.

As the more characteristic deposits of the glacial belt are due to ice, and those of the tundra belt to

[1] Richthofen, F., "On the Mode of Origin of the Loess," *Geol. Mag.* 1882, p. 293.

water, so those of the steppe regions are due to wind—they are essentially aeolian. The predominance of wind transport is due to aridity: the areas are essentially desert, for the temperature, which is here low, and in the subtropical deserts high, does not markedly affect the nature of the deposits, so far as the matrix is concerned.

The actual cause of the difference between the size of grain of the deposits found in the two kinds of areas is that the steppe region, like the tundra region, was glaciated in Pleistocene times, allowing of the occurrence at the present day of much glacial mud, which, being transported as dust by the winds, settles down on the floor of the steppes, contrasting with the prevalent sands of the subtropical zone.

There is a certain amount of vegetation in the steppes, but the aridity is such that it appears to be in the main decomposed before being covered up to form peat-layers: peat deposits are therefore not found in the typical steppe areas, thus giving a distinction between them and tundra regions. It is true that on the borders between steppe and forest, in conditions less arid than those of the typical steppes, a certain amount of humus is mingled with the dust, thus giving rise to the Tchernosem or Black Earth.

Some of the characteristics of the Loess may now be mentioned. It is noticeable that its distribution has little reference to altitude, save that on the whole it is found in intermont basins, and it retains its characters unaltered over very wide areas. The typical Loess is unstratified as opposed to aquatic loams and brick-earths, which otherwise resemble it in many respects. The particles are oriented in all directions. Alkaline salts due to evaporation of solutions are found in it. It has a vertical fracture, a false cleavage due to numerous vertical tubes formed by roots of grass-like plants growing down into it. These tubes often have a coating of carbonate of lime deposited at their margins. Certain concretions, the Loess-püppchen, are found often in considerable numbers. They are sometimes simple spherical or elliptical bodies, but are frequently compound, the components being symmetrically arranged. By taking into consideration these combined characters the Loess may be recognised, as it has been in the only example hitherto recorded in Britain, which is interbedded between two glacial deposits on the Durham coast. It was described by Trechmann in the *Quarterly Journal of the Geological Society* for 1920, p. 173.

The only other important accumulation of the steppes is *rubble-drift*, a colluvial accumulation which occurs on the slopes of the higher ground, graduating downwards into the fine material of the lowlands. As with the tundra-rubble so here the material is largely due to mechanical fracture of the rocks, followed by transport to lower levels by infrequent but violent falls of rain.

The organisms of the steppes may next be noticed. Plants are unimportant as fossils for the reasons already mentioned, which forbid their preservation to any great extent, so that we are not likely to meet with fossil plants in ancient steppe-formations. The chief animals preserved are mollusca and mammalia. The mollusca are all land forms, as *Pupilla*, *Eulota* and other genera, though *Succinea*, a swamp form, is occasionally met with where moisture has been present. True aquatic forms, like *Unio*, *Cyclas* and *Planorbis*, are entirely absent. This is an important point, militating against the theory that the Loess was formed as an aqueous deposit. The mammalia are of interest, as giving support to the view put forward on other grounds that the Loess is an ancient steppe-formation. Many of them are typical steppe-forms, such as the Saiga antelope, Prjevalski's horse, the steppe-porcupine and the jerboa. It is true that others, such as the mammoth, are found in the Loess. In the same way, however, large mammalia not characteristic of the steppe are sometimes found roaming over it in search of food in times of stress.

(4) *Temperate Forest Belt.* This once existed in its virgin state in the central part of western Europe, but cultivation has caused great changes, though portions of the primaeval forests still remain intact. The actual forest regions are marked by a mixture of residual and cumulose material constituting soils *in situ*. The mechanical residual material is usually mixed with a certain amount of humus, preserved from destruction by the humidity. The residual matter is composed of weathered rock-particles from which the soluble constituents have been largely leached out as the result of chemical weathering brought about by the rainfall. In these residues silica and aluminium silicates preponderate, the weathering being kaolinitic

and not lateritic. The aqueous deposits, whether
fluviatile or lacustrine, formed by transport and settle-
ment of the weathered matter will have a similar
character. Where forests are absent, humus still tends
to be preserved and in suitable areas of comparative
flatness and with a large supply of surface water, the
types of peat already mentioned as characteristic of
this belt will be formed, as in the Fens of eastern
England and in upland regions. This peat is of two
kinds, dependent on the nature of the surface waters.
When the water circulates freely the peat gives an
alkaline reaction, but when it is stagnant, an acid one.
The former type occurs in the East Anglian Fenland,
the latter in many of the upland peat-bogs.

Rubble-drifts are not extensively formed, as frost
action which gives rise to rubble-drifts in colder regions,
and the great contrast of diurnal and nocturnal temperatures, producing similar accumulations in desert
areas, are not greatly or widely effective.

The lacustrine deposits are marked on the one hand
by the general absence of boulders embedded in fine
sediment, a marked feature of the lakes of the glacial
and tundra belts, and on the other hand by the absence
of such chemical precipitates as characterise the
salt lakes of the subtropical desert belt.

Little need be said about the plants and animals.
Where the deposits are comparatively recent, and the
organisms therefore largely of existing species, their
determination as belonging to the forest belt is simple.
In the case of more ancient deposits, apart from the
absence of special modifications, such as occur in

steppe and desert regions, there is little to help us, and in such cases reliance must be placed chiefly on lithological characters.

(5) *The Subtropical Desert Belt.* The desert belts are on the whole more regular than those of steppe and temperate forest. That in the northern hemisphere runs almost continuously in the Old World, extending through the interior of Asia and continuing into Africa as the Sahara and its subsidiary tracts. In the New World it is much less extensive, occurring only in the western part of the United States. The southern belt, owing to deficiency of land, is much smaller. Deserts are found in Australia, South Africa and South America.

Deserts have been described under four heads, namely, rock-deserts, gravel-deserts, sand-deserts and loam-deserts, each presenting characteristic features. The rock-deserts do not concern us here, for as the name indicates, they are bare of deposits. A desert area may have a floor consisting of more than one of the varieties above named: in the case of a large desert all varieties may be found occupying different tracts.

Before describing the gravel-deserts a few words must be devoted to a consideration of the main modes of denudation. Owing to aridity wind is the main agent of transport, though water is far from negligible: for although a desert may be rainless even for years, rain falls sooner or later, often with torrential violence, especially in the upland regions. The bare rocks are weathered and disintegrated by the mechanical action of change of temperature, which produces the greatest effect. The angular detritus so broken up is largely transported from the high grounds by the intermittent freshets of the rainy periods, and is often carried some distance into the flatter parts, becoming finer away from the hills and intermingling with the blown sand. In this way dry delta-fans of coarse rubble deposit are formed, and a number of such fans may coalesce at their margins, producing a fringe of rubble deposit between the uplands and the sandy tracts of the plain. In the intervals between the formation of these rubbly layers sand may accumulate, and thus alternate layers of sand and rubble will be produced. Should the wind increase in velocity after accumulation of such alternating layers, and should such increase be maintained for some time, it will transport the sand to other regions, and the fragments of the rubble will be left behind, thus producing a gravel-desert. The fragments of the gravel undergo further change, being worn smooth by the action of the wind carrying sand-grains over them, and the sand will also impart a polish to the fragments, so that the pebbles of such a gravel-desert often have an appearance as though varnished.

Many isolated pebbles become faceted, forming the so-called Dreikanter. These are often abundant in deserts, but are not by any means always found where pebbles occur, and they are also formed under conditions other than those obtaining in deserts, so that although to some extent characteristic of deserts they are not absolutely so.

The sand-deserts are the most widespread tracts of desert regions: the sands present certain important features. The first to be noticed is their composition. It has been seen that weathering is almost entirely mechanical; accordingly the soluble constituents of the rocks are not leached out by weathering, and the constituent grains have the composition of the rock-particles from which they were derived. Therefore we find grains of quartz, felspar and ferromagnesian minerals, each retaining its original composition, a marked distinction from the grains of mechanical accumulations in regions of considerable rainfall. Many years ago, after examination of the deposits of the Nile Delta, Prof. Judd showed that as the grains had been blown into the river from the desert, and not submitted to any appreciable water action by the river, they had not undergone any leaching and retained their desert features[1].

As the accumulated sands of the deserts are often built up into dunes, with their marked false-bedded structure, this is an aid in distinguishing the desert deposits of past times.

A third and very important character is the rounding of the grains, due to greater friction between them during air-transport than when water is the principal agent. The grains are not only often nearly spherical, but also on the whole smaller than those worn by water action, and they often receive a high polish. This is an important criterion.

[1] Judd, J. W., "Report on a Series of Specimens of Deposits of the Nile Delta," *Proc. Roy. Soc.* No. 240, 1886.

M

The loam-deserts are those which are due to deposition of material in playa lakes, that is, sheets of
water that form temporarily in many deserts, such
as the Schotts of North Africa. Mechanical sediment,
sand or dust, is blown or washed into them, giving rise
to laminated deposits, but as the result of periodical
evaporation, various salts are deposited, giving important indications of desert conditions. The water
tracts which have at present no outlet to the sea
derive their waters from two sources: in one case they
are derived solely from rivers. Here the precipitates
are largely carbonates and sulphates; in the other case
they are tracts severed from the sea, and the precipitates consist of salts found in sea-water, including
much sodium chloride. These chemical precipitates
give very important indications of the prevalence of
desert conditions in past times. As regards the various
markings and structures characteristic of the dried
surfaces of the playas (rain-prints, sun-cracks, salt-
pseudomorphs, etc.), they have been referred to in a
previous section.

Lithology, then, furnishes us with very important
criteria for ascertaining the desert origin of deposits
of the past.

The organisms of the desert accumulations are not
extensively found in the older deposits, save in those
formed in water tracts separated from the open ocean.
As before stated, plants are specially modified to resist
the arid conditions, but the same conditions usually
forbid their preservation, save where carried into
playa lakes. Terrestrial animals are not abundantly

found, and do not necessarily give indications of desert conditions. Various amphibian and reptilian remains occur in British Permian and Triassic rocks of desert origin. Among the Invertebrata scorpion-like animals may be expected, and are in fact found, but scorpions, though frequent in deserts, are by no means characteristic.

The evaporating waters of desert lakes formed by severance from the open ocean give us important information. As the water becomes salter and accordingly less suitable to normal marine life the species are gradually extinguished, leaving only the more hardy, which themselves will disappear as complete desiccation approaches. Accordingly, though individuals may be numerous, species will be few as compared with the open ocean. The molluscan tests become modified under the adverse conditions. A study of changes in the shells of the cockle (*Cardium*) in the beaches left at various heights above the receding waters of lakes of the Aralo-Caspian area was made by W. Bateson[1]. The highest beaches were formed when the waters had undergone least evaporation and the shells in these show least change. The changes are various: the shells become smaller, less symmetrical and thinner. Similar modifications may be found in the shells of some of the mollusca of the Permian Magnesian Limestone of the north of England.

One deposit of a cumulose character is found in

[1] Bateson, W., "On some variations of *Cardium edule*, apparently correlated to the conditions of life," *Phil. Trans. Roy. Soc.* Ser. B, vol. 180, 1889, p. 297.

exceptional circumstances in the desert belt, namely, guano. It occurs on islands and seaboards of the mainland in practically rainless areas, as very slight rainfall will cause its destruction. It is therefore unknown among the ancient sediments, although it may occur in Pleistocene caverns as it certainly does in some modern ones. Soluble as the material is it may yet be preserved by being carried down in solution into the rocks on which it was deposited, causing phosphatisation by metasomatic action. As modern instances the phosphatised coral limestone found by Sir J. Murray in Christmas Island, and the phosphatised trachyte of Clipperton Atoll described by Sir J. J. H. Teall may be cited. Similar cases of phosphatisation may eventually be found among older rocks.

(6) *The Tropical Forest Belt.* This comprises the humid regions found in southern Asia and central Africa in the Old World, and in Brazil and adjacent countries in the New. It is marked by luxuriant forest growth, favoured by the great heat and abundant precipitation of rain. The conditions are remarkably uniform over wide areas. The accumulations are largely sedentary, both residual and cumulose. A very important feature of the residual material is due to the mode of weathering, spoken of as *lateritic.* It was seen that in temperate forest regions though weathering is largely chemical the residues contain silica in chemical combination with alumina. The result of lateritic weathering is to separate the silica from the alumina, each occurring in the free state. The lateritic

soils also contain a considerable amount of ferruginous matter, and occasionally a certain amount of carbonate of lime: this variety of laterite is called kunkar in India. The real reason for the separation of the silica from the alumina has not been definitely determined: it is sometimes regarded as due to bacterial action[1], lateritisation being a sort of disease.

Owing to the abundant vegetation of the belt humus is usually found in considerable quantity in the accumulations, but owing to the great heat much of the vegetation is completely decomposed before being covered by fresh material; nevertheless nearly pure cumulose deposits, comparable with the peat of higher latitudes, are more common than was formerly supposed.

Conditions favourable for this formation are specially prone to occur along the tracts bordering the sea and large rivers, where carbonaceous deposits of some importance occur in the flatter tracts, often in association with mangrove swamps.

Little need be said of the organisms of the tropical forest belt. Plants may be preserved in river alluvia and lake deposits, and if closely allied to existing forms give indications of tropical conditions. The fan-palms, feather-palms, aralias and other plants of the European Eocene strata point to such conditions having been prevalent during this period. The animals to some extent give similar evidence, though in less degree.

[1] Holland, Sir T. H., "On the Constitution, Origin and Dehydration of Laterite," *Geol. Mag.* 1903, p. 59.

From what has been said it will be gathered that by combining the evidence from lithology and from organic contents, we usually obtain sufficient data to enable us to determine the climatic conditions under which the deposits of past times were formed, even when of great geological antiquity. It will be noticed that much more information on this matter is supplied by lithology than by organic contents.

Shifting of the Belts in Time

The geologist has long been aware that any particular area, Great Britain for example, has undergone marked vicissitudes of climate in past ages. As examples, a warm arid climate prevailed in parts at any rate of the Triassic period, a warm humid climate in parts of the early Tertiary epoch, and a glacial climate during a portion of the Pleistocene period. Geologists have perhaps been too apt to assume that if they obtained indications of, say, a warm cycle during part of a long period, similar conditions probably prevailed during the whole of it. The Silurian period of N.W. Europe has often been spoken of as a warm one, from evidence which will be considered later, and so far as parts of it are concerned, the inference is almost certainly justified, but it does not follow that the whole period was necessarily marked by warmth in that area. With marine deposits, it is not always easy to establish the climatic conditions under which each of the subdivisions was laid down, hence the importance of the deposits of the land areas in furnishing evidence as to epicycles.

This may be illustrated by reference to Tertiary times. It was long held that in our country a cycle of warm climate marked this period until the Pliocene, and was then succeeded by a short cycle of cold, culminating in Pleistocene times. So far as cycles are concerned this is true, but the study of the terrestrial deposits indicates that the cold cycle was marked by epicycles, of alternating cold and warmth, but could we get similar evidence in the case of the early Tertiary rocks we should doubtless detect epicycles there. The cold Pleistocene period was once believed to be one ushered in by comparative coolness, culminating in one glaciation, and ending in increase of temperature causing the waning of the ice. Recent research tends to show that the changes were not so simple. During the cycle of cold several alternating epicycles of glacial and interglacial conditions are now believed to have occurred, and it is by no means clear that the cold cycle has terminated: we may be actually living in an interglacial epicycle of this cycle, and at no distant geological date our country may be again covered by a mantle of snow and ice.

Nehring, in a paper previously cited, indicated that as regards the climatic conditions of the glacial, tundra, steppe and temperate forest belts there was a distribution in time as well as in space, and of recent years more evidence of this has been obtained. Beginning with a temperate period in any area, there might be a sequence in later times to steppe, tundra and glacial conditions, followed by a sequence in the reverse order. If there were two glaciations, the interglacial period between them would be marked at first by change from glacial towards temperate forest and later in reverse order towards glacial. The milder stages need not of course be reached. Furthermore, as in space, so in time: a glacial episode might be succeeded by a steppe, or a tundra by a forest, without the intervening stages. Some such sequence has been established in the case of the Pliocene, Pleistocene, and Holocene deposits of N.W. Europe, but the task is not yet completed.

The above illustration is furnished by events happening in a cold cycle. During a warm one, there might be epicycles marked by changes from tropical forest through desert to temperate forest and vice versa.

The considerations set forth above will show the high importance of the study of the deposits of the land masses, as throwing light on certain geological problems, which cannot at present be solved by the study of marine formations alone.

CHAPTER VI

MARINE DEPOSITS: BELTS OF SEDIMENTATION

Horizontal Changes. On a large scale the deposits of the ocean are arranged in definite belts. One set is determined by climatic conditions: these are analogous to those described in the last chapter, dealing with accumulation in the land areas, and like them running parallel to the equator. They are marked rather by variations in the floras and faunas than by lithological changes, though the latter are not negligible. These belts will be considered in a later chapter. Those now to be discussed are largely determined by distance from the coast-line, though partly by variations in the depth of the water. The belts run concentrically around the great abysses of the ocean, that nearest the land being limited by the coast-line. These belts were largely traced during the explorations conducted on the *Challenger* expedition, and recorded by Murray and Renard in the volume describing the specimens of deep-sea deposits collected on that expedition. From the geological point of view some modification in the classification will be found useful. The main subdivision into terrigenous and pelagic is of prime importance.

In the terrigenous belts are accumulated mechanical deposits formed essentially of material produced by

denudation of the land masses, though some of it is
of direct volcanic origin. There is also a considerable
amount of deposit of organic origin in the terrigenous
belts. Outside these belts are the organic deposits,
both calcareous and siliceous. Where these are confined to the oceanic abysses, they constitute the
organic oozes, but in places the terrigenous deposits
are succeeded outwards by organic deposits which are
not abyssal. Such shallower water deposits outside
the terrigenous belts are of considerable importance
to geologists, more so than the organic oozes. Lastly,
in the extreme abysses of the deep sea "red clay" is
met with, so that the belts of the other sediments may
be regarded as arranged in a roughly concentric
manner around the abyssal clay.

The line separating the outermost terrigenous deposits from the organic deposits beyond was termed
by Murray the *mud line*. It is of course not a mathe-
matical line, but a belt of varying width, in which a
gradual transition from mud to organic material
occurs. But to the geologist there is another important line on the inside of the mud belt, which
might equally well be termed the mud line.

It is usually stated that the mechanical deposits are
laid down in the order of size and weight of their
constituents, the largest and heaviest being dropped
on to the ocean floor nearest the land, the finer
material further away. This is true on a small scale,
but on a large scale there are so many exceptions
that the terrigenous deposits on the shoreward side of
the mud belt cannot be divided into minor parallel

belts by reference to variations in the size of their
constituent fragments. On the contrary, the sea-floor
of this inner belt is often marked by stretches of
pebbles, sand, silt and mud, or even organic deposits,
lying side by side with no very definite arrangement.
Such a varied series of deposits is well displayed in

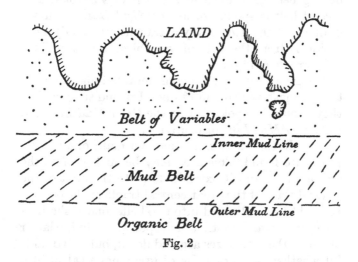

Fig. 2

the Irish Sea, as may be gathered from a perusal of
the Report on the "Marine Zoology, Botany and
Geology of the Irish Sea" by a committee of the
British Association (*Report*, 1895), in which the main
results are summarised.

Accordingly this inner belt adjoining the coast will
be spoken of henceforth as the *Belt of Variables*. (See
Fig. 2.) Outside this the conditions become more

uniform. Open ocean is as a rule met with; soundings
are more uniform, though on the whole steadily
deepening away from the coast; the currents are more
regular in direction and velocity. The combined result
is greater uniformity in the character of the sediment,
which is almost entirely mud, the coarser material
having been laid down in the belt of variables. The
second belt is spoken of as the *Mud Belt*. Its inner
limit is a line (often really a transitional band) separating it from the belt of variables: its outer limit
is Murray's mud line.

The inner line, of very considerable importance to
the geologist, is also a mud line. I would suggest that
they should be spoken of as the *Outer Mud Line* and
the *Inner Mud Line* respectively.

Outside the outer mud line the main organic deposits are found. This area may be spoken of as the
Organic Belt, and it extends down to the limit of the
abyssal clays of the great central tracts of the oceans.
It might be subdivided into two belts, one of shallow-
water tracts, and another of deeper water, in the latter
of which the true oozes are laid down, but it is doubtful whether this would be of great practical utility,
for in a large number of cases, at all events at the
present day, the mud belt is immediately succeeded
by oozes, as the muds extend outwards into the
abyssal regions. Nevertheless it is important that the
geologist should recognise that organic deposits not
coming under the definition of oozes are being formed
in places outside the outer mud line, and there is
evidence that similar deposits have been so formed in
past times.

The width of the different belts depends on more than one circumstance. As a general rule the belt of variables extends only a few miles from the coast, if so far, while the mud belt is as a rule far wider, for it was found during the *Challenger* expedition that the mud usually extended 100 miles from the land and in exceptional cases even 200 miles. In other cases, likewise exceptional, it is entirely absent.

Variations in depth play a very important part in determining the distribution of the component sediments of the different belts. The existing oceans for the most part present cross-sections resembling that of a tea tray. Just off the coasts are shallow waters sloping on the whole seaward: these form the *continental shelf*. From this shelf a slope of varying steepness leads down to the abyssal parts of the ocean. The belt of variables is mainly confined to the continental shelf but not always so, and the exceptions, as will be shown, are of considerable importance.

The mud lines are not determined by depth. The two lines that are so determined are first those separating the shallower part of the organic belt from the deposits of organic ooze, and secondly the oozes from the red clay. Neither of these concerns the geologist to the same extent as the mud lines, for the deep-sea oozes are not extensively represented among the rocks of the past, and it is doubtful whether the deep-sea clay has any such representative. This is due to the smaller likelihood of deposits formed at great depths being brought up by the earth-movements to form constituent parts of the land. Consequently the strata with which the geologist is chiefly concerned are those laid down in the belt of variables, the mud belt, and in a less degree in the organic belt.

Vertical Changes. At any one time the belts and their divisional lines will appear as in Fig. 3. But in course of time the belts tend to overlap one another, with a corresponding shifting of the lines of separation. On a small scale this shifting is due to more than one cause, but on a large scale it is due to changes in the position of the land margins. With an advancing coast-line shallow-water sediments overlap those laid down further out to sea; when the coast is receding the opposite will occur. The advance of the coast-line may be due either to emergence of the land by earth-movements or to the silting up of the shallows by deposition of sediments, converting the shallow sea into land, while the recession of the coast and consequent encroachment of the sea may be due to submergence by earth-movement or to erosion. It is clear that in the case of advance of the coast-line both emergence and silting may occur together, and similarly in the case of retreat subsidence and erosion may be concurrent. The effects of earth-movement are no doubt more important than those of silting or erosion.

For simplicity's sake let us consider a shallow-water tract where the land is gradually encroaching upon the sea. (See Fig. 3.) A stratum of coarse mechanical sediment will under ordinary conditions be laid down next the coast-line, a; further out, finer sediment will be simultaneously deposited as mud, b; and still further away there may be organic sediment, c. If the first

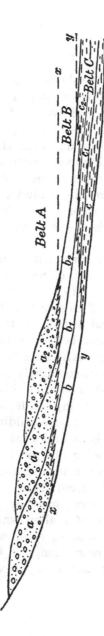

Fig. 3

a. Coarse mechanical sediment.
b. Fine mechanical sediment.
c. Organic sediment.

stratum of sand silts up the shallow water, turning it into land, the next stratum of sand, a_1, will be deposited further out, and will overlap b, and b_1 will overlap c, and so on. As this process goes on it will ultimately result in the arrangement shown in the figure. Hence there are three continuous sets of strata superposed one over the other, for instance, the lowest of organic material, the middle of mud and the upper of variable material, or on a smaller scale, the lowest, say, of mud, the middle of sand and the upper of conglomerate.

According to generally prevalent ideas the lowest mass is the oldest and the highest the youngest. Although this is true for a set of deposits met with in a vertical bore-hole, it will be seen that it is not true over a considerable lateral stretch. A stratum far out from the original coast-line belonging to the lowest series may be newer than the first-formed stratum, a, of the uppermost deposits: accordingly, as observed in Chapter II, the contemporaneity of a lithological belt is not actually proved by tracing the deposit laterally. We are dealing with a kind of false-bedding on a large scale, the divisional line between two strata of one of the lithological bands being oblique to the line of demarcation between one band and another, which is not a true line of stratification. It may be remarked that this may well be accompanied by contemporaneous erosion in the case of shallow-water deposits. Though the true strata, a, a_1, a_2, etc. are not exactly parallel to the lines of demarcation, x, y, between two lithological bands, they may be approximately so;

accordingly it may be difficult to distinguish their actual non-parallelism in the field. This non-parallelism is greater in the case of the coarser sediments, less in the finer. Hence the lines separating band from band are more nearly coincident with time-lines in the case of the finer sediments. The actual time-lines run through different kinds of sediment. Thus in Fig. 3 a time-line separates the deposits a, b, c from a_1, b_1, c_1, and another separates the latter from a_2, b_2, c_2 and so on.

Another interesting effect of the above-described mode of deposition is connected with measurement of thicknesses of strata. This may be done in two ways, and in many cases it is difficult, if not impossible, to tell which way has been adopted. In the one way, true vertical thicknesses are obtained by measuring at right angles to the lines separating one lithological band from another. In the other, the thickness of the strata of any one band is obtained by measuring the actual beds between the planes of stratification. Thus in Fig. 3 the thickness of the A deposits for instance would be obtained by measuring each individual stratum a, a_1, a_2, ... and adding up the total, whereas the actual vertical thickness is found by measuring a line at right angles to the junction of the a and b bands. The former mode of measurement might give a thickness of thousands of feet, the latter of only a few score. The discrepancy is most marked in the case of shallow-water sediments, such as are laid down as deltas and sandbanks. It has often been noted with surprise that thousands

M   6

of feet of sandstone have been laid down in water that remained constantly shallow. If the measurement was of vertical thickness, it would imply the subsidence of the sea-floor to a depth commensurate with the thickness of the deposit, for the area must begin as shallow water for the deposition of the first-formed shallow-water sediment, and must continue shallow throughout, for no sediment can be accumulated if the water tract be completely silted up. If however the actual vertical thickness be but a fraction of that

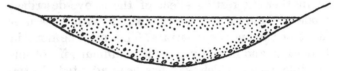

Fig. 4. *Cycle in a Gulf.*
Dotted = coastal deposits.
Plain = open-water deposits.

obtained by adding together the sum of the measurements of the individual strata, the amount of subsidence necessary is very much less and the difficulty disappears.

Cycles and Epicycles of Deposition. We are now in a position to consider the processes of sedimentation during cycles and epicycles of deposit, beginning with the cycle, which is the interval between two terrestrial phases, when the sea encroaches on and occupies the area. It may be most simply illustrated by taking the case of a gulf-like tract of comparatively limited size before considering what happens in the case of the open ocean. (See Fig. 4.)

Starting with a low-lying tract, nearly surrounded by uplands, the water will first invade the lowest part, where sedimentation will start, and as subsidence proceeds larger and larger tracts will become sea and the coast-lines will gradually recede from one another: simultaneously the area receiving sediment will be enlarged. The first-formed sediments will be exclusively coastal, as the shores are then not far from the centre of the water area, but as the water spreads the time will arrive when the coarser material cannot be carried out to the centre of the gulf, where more open-water sediment will be laid down. This will go on so long as the coasts recede, other things being equal. The time will arrive, however, when either by emergence, or by conversion of shallow water into land by silting, the coasts will once more begin to converge, and the shallow-water deposits will then begin to overlap the open-water ones, and may finally spread over the whole gulf as shown in the figure.

In the great ocean tracts it is most improbable that the whole area will be filled with coastal sediments. Also, except in very early geological times, it is doubtful whether a great ocean has ever come into being; hence, shallow-water deposits will not be found over the central parts of the ocean tracts, even at the beginning or end of a cycle.

It is generally supposed that alternate submergence and emergence during long geological ages only affected what are called the critical areas, that is, those which at the present day are occupied by the continental shelf, below water level, and the coastal plain above it,

the cores of the continents and the central tracts of
the oceans remaining permanently as land and sea
respectively. It must be understood, however, that
these cores and central deeps may undergo change of
size and position in course of time.

During a period of maximum emergence, the critical
area would be land, a coastal plain; during one of
maximum submergence, it would be sea, a continental shelf; at intermediate times, it would be partly
coastal plain, partly continental shelf. This is largely
the existing state of affairs.

During a marine cycle, the belts of open-water
deposit would form a wedge with its apex pointing
landward: while during a land cycle, preceding or
succeeding a marine cycle, the coastal deposits would
form a wedge with its apex pointing seaward. (See
Fig. 4, which shows two wedges formed during land
cycles, with an intermediate one, pointing the other
way, formed during a marine cycle.) The section shown
in the figure is purely diagrammatic: actually, complications would arise owing to the production of an
unconformity during each period of emergence, so
that the wedges on the land side would be partly
destroyed by erosion, though they would remain complete outside the critical area on the seaward side.
(See Fig. 5.) The approach to simplicity would be
greatest when the final conversion of the erstwhile sea
into land was mainly brought about, not by emergence,
but by silting up of the shallow waters by accumulation of sediments.

It results from the foregoing that each marine cycle

begins with a shallow-water phase, is ultimately succeeded by a more open-water phase, if the area is of
sufficient extent to prevent the transport of coarser
material to the centre, and ends with another shallow-
water phase. Accordingly, in a complete marine cycle,
the lowest deposits belong to the belt of variables,
those above to the mud belt, and this again may be

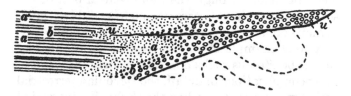

Fig. 5. Showing effect of alternate submergence and emergence
on the distribution of sediments. *a*. Deposits formed during period
of submergence forming a wedge with apex of the more open-water
sediments pointing landward. *a'*. Lower half of a larger wedge
formed during a later period of submergence. *b*. A wedge formed
during emergence with coarse sediment having apex pointing
seaward. *b'*. Upper half of an earlier wedge of similar origin.
u—u. Unconformity where actual conversion of sea-floor into
land occurred. The line of unconformity points to the apex of the
wedge *b*.

succeeded by the organic belt, granted sufficient time
and space. When the coast-lines cease to recede and
begin to approach one another, the sequence is reversed, organic deposits being succeeded by muds and
these finally by variables. It follows that there is a
tendency for one belt to be succeeded by another
most nearly allied to it as regards distance of origin
from the coast, a variable being succeeded or preceded

by a mud belt, and the latter by an organic one, but not an organic by a variable, though this may and does occur exceptionally. Further it follows that, if we could obtain a vertical section through the deposits formed in one marine cycle near the margin of a water area, we should find only deposits of littoral character, perhaps variables only, whereas if a bore could be made through the more central part of the sea tract, we should pass through variables, mud, perhaps organic, then mud, and lastly a variable belt.

A good example of such a sequence is afforded by the Carboniferous rocks of southern Britain. Eliminating complications of detail, the Devonian terrestrial period was succeeded, when the Carboniferous period of submergence began in the Bristol and South Wales area, by conglomerates and other deposits of the variable belt, then the Lower Limestone Shales, forming the mud belt, the Mountain Limestone representing the organic belt, the Upper Limestone Shales the second mud belt, and the Millstone Grit and Coal-measures the upper belt of variables. It is noticeable that in Devonshire, which is near to the ancient coast, the organic belt is absent. The sequence of deposits in the more open sea of this period and the disappearance of the organic belt towards the margin are also shown by the Carboniferous rocks of the north of England. It would appear from the great difference in thickness between the deposits of the upper and lower variable belts over most of the British area that the initial shallow-water phase of the cycle was very brief as compared with the final phase, though this may be partly due to another cause, which will be discussed later, namely, the rapid accumulation of sediments of deltaic character.

Epicycles. That these occur is made manifest in various ways, but especially by changes in the characters of the deposits when examined in vertical succession. During a marine cycle consisting of the three phases above noted, shallow, deep and shallow water successively, conditions may remain fairly uniform during any one phase, but almost invariably there are departures from uniformity, marked by alternations of sediments formed under different conditions. These denote epicycles. Thus thick masses of sandstone are often found to be interstratified with bands of shale, frequently with almost rhythmical regularity, and similarly we find bands of limestone intercalated in muds. The Millstone Grit of the north of England furnishes a good example of the former and parts of the Lias of the latter. The alternations may be due to more than one cause, and it is not always easy to detect the actual one. Shifting in the direction and change in the velocity of currents is no doubt operative at times. Another cause is climatic change. That such is effective on a small scale has been shown by Baron de Geer in the case of the varve clays, where the alternations are due to seasonal change, and are therefore exceptionally regular. Variations in the nature of sediment may occur on a larger scale, as the result of greater climatic changes spread over long periods, recurring with some regularity, though of course not comparable with the time intervals of the seasonal changes. The matter requires further study and light may be thrown on it by comparison of faunas and floras. But important as the causes just described are, a large number of the alternations in the nature of sediments must be due to minor emergences and submergences during one cycle. Shale beds in sandstone would thus be due to minor periods of submergence; sandstones in shales to minor periods of emergence; and so with other intercalations. Such changes do not necessarily imply negative movement; they can be brought about by positive movement only, in this case depending on the ratio between rate of movement and that of sedimentation. If subsidence is more rapid than deposit, the coasts of the water tract will recede. If deposit is more rapid than subsidence, the coasts will converge by silting up, and sands will be deposited over sites formerly receiving mud. This will be more marked if there be an actual pause in the movement. It is quite possible in this way for the whole of the deposits of a cycle to be formed during a period of positive movement only, and this appears to have been approached in the case of the British Carboniferous rocks, though the minor complications are undoubtedly due to negative movement.

CHAPTER VII

MARINE DEPOSITS (*continued*):
THE BELT OF VARIABLES

Characteristic Features. At the present day the most characteristic features of the belt of variables are displayed by deposits accumulating upon the continental shelf. The deposits of the shelf are not necessarily of very great thickness as compared with those further away from the coast-line, but they have a wide areal distribution. The actual existence of a shelf is not necessary for the formation of the kind of deposit now being considered. The essential features of the deposit are primarily due to the fact that the sea-currents, whether tidal or due to other causes, are in action down to the sea-floor. It is not apparently agreed as to how far down currents produce an appreciable effect in disturbing the sea-floor. The depth of course varies with the magnitude of the currents. They appear to be effective to a depth of 100 fathoms, and as this is about the depth of the outer edge of the continental shelf on which the British Isles stand, we may take it that the currents do definitely affect the sea-floor of that region, while off the edge, the depth increases so rapidly that the sea-floor is practically undisturbed by currents not far out from the edge of the shelf. As the upper layers of the sea-water are still disturbed by superficial currents, finer mechanical sediment is carried over the site and deposited; this area of deposition extends to the outer mud line. Usually mud alone is extensively deposited beyond the shelf, and the inner mud line approximately coincides with its edge, but as will be seen presently there are some conditions under which coarser sediment is laid down beyond the shelf, when the inner mud line will extend beyond it, and the belt of variables will encroach upon the tracts of deeper water.

We are for the present concerned with the deposits on the continental shelf, where currents operate right down to the bottom. One feature of this tract is the great variability of the currents, as regards direction and velocity, with frequent production of eddies. These variations occur laterally at any one time, and at different times in any one spot. One important consequence is the great difference in the nature of the sediments formed contemporaneously in different parts of the shallows, and their frequent rapid change in a vertical direction. Anyone conversant with the shallows around the coast knows how rapidly the sediments change laterally, both in passing out from the land seawards and in proceeding parallel with the coast-line. Thus a common sequence from the coast-line outwards is from pebble-beaches to sands and then to muds. There are, however, many important exceptions to this, which will be discussed later.

Very important are the lateral changes displayed along a tract subparallel with the coast-line. For instance, instead of meeting with a continuous belt of sand only, as we might expect, we may find extensive patches, here of mud, there of organic material, due to a variety of local causes, such as the entrance of a river, which would cause deposition of mud, or the accumulation of a host of dead shells, giving rise to a calcareous deposit.

Two structures, though also found in terrestrial deposits, are, in the case of marine sediments, confined to those formed only where currents operate on the sea-floor. These are ripple-marks and false-bedding. They are mainly confined to the belt of variables, though exceptionally formed in the mud belt when this extends into shallow water.

The action of currents on the floor of the continental shelf has important effects. If they are sufficiently powerful, the floor is kept free from sediment, and bare rock is exposed, as on the wave-cut platform. This is often fissured owing to erosion along joint-planes, and in these fissures sediment may be entrapped and permanently preserved. Such occurrences are known amongst the older rocks. When sediment is laid down on the continental shelf, there is an important difference between its method of accumulation there and its deposition in deeper water, where the floor is not affected by current action. Here material once laid down is not disturbed unless changes occur which raise the floor to the level where currents can affect it. In the shallows, on the contrary, actions take place which cause the removal of some of the sediment shortly after its deposition, and accordingly the amount of sediment finally preserved may be but a tithe of that which has from time to time been deposited.

The removal may take place in more than one way:
after a certain amount has been deposited, increase
in the velocity of the currents may cause the removal
of the upper part, leaving the lower intact. Subsequently fresh material may be laid down, producing
a little unconformity, and this process may be repeated
again and again, giving rise to false-bedding, often on
a large scale. The removal of parts of the deposit may
however go on concurrently with the deposition, without the production of any break. The deposits as first
formed may not consist of particles of uniform size
and weight, for various reasons. Gentle agitation of
the water, with much eddying action, will often raise
parts of the sediment above the floor, and while thus
suspended the finer particles may be carried elsewhere,
while the coarser ones sink again to the floor in the
same place. As this action continues the coarser
particles themselves are reduced in size, and the finer
chips, whether thus produced or existing previously,
will by degrees be carried into deeper water, and
ultimately beyond the belt of variables into the mud
belt. By these winnowing processes, therefore, the
amount of sediment permanently deposited in the belt
of variables will be much less than it would otherwise
be, and a very long period of time may be represented
by deposits of no great thickness, and yet having no
important physical breaks occurring in them, but
only a large number of very small interruptions, of
which no one is marked by any important change in
the nature of the fauna. To illustrate this we may cite
as an example the Eocene rocks of southern England.

These deposits, with a maximum thickness of a few hundred feet, were once regarded as having been formed in a short time, and there is no important palæontological break therein. Nevertheless they are represented elsewhere by many thousands of feet of open-water sediment, namely, the Nummulitic Limestones and associated strata, obviously requiring a long period of time for their deposition.

Some details regarding the process of winnowing may now be given. It occurs frequently in the case of ordinary mechanical sediment of different degrees of coarseness, and it is quite common that the material left behind is extraneous, such as shells and glauconitic and phosphatic nodules. Any of these may be concentrated in a deposit by the removal of the finer sand and mud, and accordingly the deposits rich in glauconite are often the result of the concentration of the glauconite by winnowing. The same is the case with many phosphatic deposits, and it is interesting to note that phosphatic nodules frequently accompany glauconitic accumulations, as in the Cambridge Greensand and in the Lower Palæozoic deposits of Esthonia. Concentration of heavy minerals must occur during winnowing, and is known on a large scale, as for example the iron-sands of the New Zealand coast, and the monazite-beaches of Brazil and Travancore. Much progress has already been made in the study of these concentrates, and the subject offers a promising field for further research.

The case of the formation of calcareous deposits by concentration requires special notice. The importance of this was emphasised by P. Lake[1], who observed a limestone in the process of formation in the Cam valley below Cambridge. Though this was a fresh-water formation, it illustrates what must occur in shallow seas. Briefly the process is as follows: a fine mechanical sediment containing shells has the matrix removed as the result of increase of velocity of the current, thus converting an essentially mechanical deposit into an organic one—a limestone. In the case of the Cam deposit both shells and matrix were moved, but to different distances; it is not necessary, however, for the larger components to be moved at all. An ancient marine limestone probably having this origin is found at the base of the Silurian rocks of West Yorkshire: it is known as the *Phacops elegans* lime-stone from the abundance of that trilobite. The bed is only four inches thick, and is practically composed of trilobite remains with little or no matrix. It is repre-sented by deposits of greater thickness elsewhere, and the simplest explanation is that the trilobites were embedded in a fine deposit, of which the matrix was washed away, leaving the tests as residue. A point in favour of this is that the trilobites are all fragmentary, consisting of heads, tails and isolated body-rings. Where similar trilobites are found in contempora-neous mud in neighbouring places they are often complete, the tests having been covered up before the animal matter connecting the segments had decayed. Many limestones of the zone of variables must have been formed in this manner.

[1] Lake, P., *Proc. Camb. Phil. Soc.* vol. XIX, 1918, p. 157.

The process of winnowing on the shallow sea-floor may be compared in a general way with those concerned in the production of residual material of the regolith on land, as the result of weathering and transport. This does not as a rule comprise the whole of the weathered material, but is the result of the excess of weathering over transport. Similarly the material deposited on the continental shelf is the excess of that furnished to the area by transport over what is removed from it by the same agent. The rest is transported elsewhere.

So long as the continental shelf is not depressed by earth-movements, deposits can be formed of a limited thickness only, corresponding with the depth of any part of the shelf below the surface of the water, which we have assumed to be about 100 fathoms at the most. No doubt in many cases the floor did sink, but in others it may have been stationary for long periods. In the former case, the thickness of the deposits may greatly exceed 600 feet.

Hitherto we have considered shallow-water deposits laid down on fairly level tracts. It is important to notice what occurs when deposition takes place on slopes of comparative steepness. These are particularly found at the fronts of deltas and at the outer margin of the continental shelf. On these slopes the mode of deposition is in a way peculiar, resembling the usual mode of construction of a railway embankment by the tipping of material over the advancing end. In this case stratification is often visible, owing to intermittent tipping. The result is that "strata" are formed parallel with the terminal slope and the deposits grow outward and not upward. As has been pointed out, therefore, the height of the embankment is small as compared with the thickness of the component "strata," measured at right angles to their planes of deposition. So it is with the natural deposits formed on steep slopes, which are actual examples of embankment formation.

We may consider first the case of the outer margin of the continental shelf, or, if this be absent, the steep slope off the coast-line. When the shelf is broad, the coarse sediment is mainly laid down on that shelf and the inner mud line may approximately coincide with the slope, the mud belt itself being in deep water. If however the shelf be narrow, or is becoming narrow by conversion of its inner margin into land, the coarser sediments may be carried out and tipped over the edge, for it is the velocity of the currents and not the depth of the water that determines the distance to which the material is transported. Under the conditions mentioned, therefore, the deposits of the belt of variables will not be confined to the shelf, but extend into the deeper waters, building out into them. As the result of this, the deep waters may be rapidly silted up by the coarse sediment and so the floor raised to the level of the shelf. It follows that beyond the shelf we may meet with thick accumulations of coarse mechanical sediment, contemporaneous with the much thinner deposits formed nearby on the shelf itself.

The Lower Ludlow rocks of the Lake District of the north of England are an example of this mode of accumulation and illustrate an interesting point. Their thickness has been measured as approximating to 10,000 feet, though this is probably an over-estimate, but in any case it is very great, too great to have been formed in a sea initially of that depth, which would approach the abyssal: the organisms throughout indicate shallow-water conditions. There must therefore have been subsidence during their formation. Now the contemporaneous shallow-water deposits of the

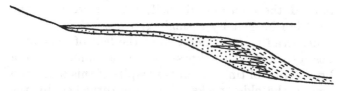

Fig. 6. Thin variables on the continental shelf and thick accumulation of mechanical variables off the shelf, when latter is subsiding and former stationary, e.g. Lower Ludlows of North Wales and the Borderland.

continental shelf show that no great subsidence occurred there. It would seem that the slope between the shelf and the outside deeps was a hinge against which marked movement of depression ceased. It is an interesting question whether such depression was due to the actual weight of the great mass of sediment as suggested by Sir J. Herschel[1]. The question is too abstruse to be discussed here.

We may now pass to delta-formation. At first deltas

[1] Herschel, Sir J., *Proc. Geol. Soc.* vol. II, 1837, pp. 548 and 596.

will be built up as embankments on the continental
shelf, if this be present, but may eventually extend
to the margin of the shelf and the process will be
continued in the deeper water beyond, repeating the
process described above. The united delta of the
Ganges and Brahmaputra is an example of one which
has grown beyond the continental shelf. We may have
coalescent deltas of several rivers, such as are found
at the present day on the north-west shores of the
Adriatic. Such coalescent deltas may also grow out
beyond the continental shelf, when instead of the
local fan formed by a single river, we should get a
widespread tract of deposits of the belt of variables,
possibly of great thickness and yet accumulated in a
brief period of time. Many examples of this are found
among the older rocks. A case is furnished by the
Carboniferous rocks of Great Britain, where deltaic
accumulations grew out into the water tract previously
receiving sediments belonging to the mud and organic
belts, though these were probably formed in a sea
of no great depth. We meet with delta growths in
Northumberland in Lower Carboniferous times, and
they spread over the whole water tract in the Upper
Carboniferous, forming the well-known sediments of
Millstone Grit and Coal-measure facies. This mode of
origin probably accounts in the main for their great
thickness, compared with the deposits of the shallow-
water phase in most parts of Britain, where delta
growth was absent, though as before suggested, short
duration of the period of formation of the latter
is largely responsible for the difference. Another

probable case of deltaic growth is seen in certain Cretaceous rocks of Saxony and the adjacent countries. The Chalk of areas further to the north-west is partly replaced by mechanical sediments, of which the Quadersandstein forms the dominant type. Still more striking is the case of the thick mass of Eocene sediments of Burma.

Before describing in greater detail the characters of the sediments of the belt of variables, a few general remarks may be made about their organic contents. Freshwater and even terrestrial organisms, as before stated, are occasionally carried into the sea, though they usually occur in marine deposits in very small proportion, while often quite abundant in estuarine deposits. Land-plants, mollusca and mammalia are not uncommon in the shallow-water deposits of the belt of variables, but are usually found in a fragmentary condition. This latter fact may be illustrated by the Eurypterids of the Palæozoic rocks. There has been much discussion as to whether they were inhabitants of fresh water or of the sea, and they are certainly found in marine deposits, as in some of the Ludlow rocks of Britain and Bohemia, which contain marine fossils. Dr Marjorie O'Connell[1], the latest contributor to the subject, has argued that they were solely inhabitants of fresh water, and that those found in marine deposits were drifted into the sea, being rare and always found in a fragmentary condition.

[1] O'Connell, Dr M., "The Habitat of the Eurypterida," *Bull. Buffalo Soc. Nat. Hist.* vol. II, No. 3, 1916, pp. 1–277.

The true denizens of the ocean, whose remains are found in the belt of variables, are mainly benthonic, for planktonic forms are not as a general rule abundant in the seas near the coast, probably on account of the character of the waters, frequently sullied as they are with fine sediment. Nevertheless, as noted in a former chapter, planktonic forms are occasionally drifted to these regions in a living or dead state, and therefore occur sparsely in the deposits. This is specially true of the smaller forms, such as the Foraminifera.

In comparatively modern deposits the relative depth at which the various sediments were formed may be judged from the enclosed organisms. Thus, among the Pliocene deposits of East Anglia, the Norwich Crag was formed in the Littoral zone, the Coralline Crag in the Coralline zone. It must be remembered, however, that dead shells are often washed from a locality, where their proper zone exists, to others where the zone is different, and especially from a deeper zone to a shallower one. Thus we find in the Littoral zone of the present day not only the shells characteristic of that zone, but others, such as *Trochus* and even *Dentalium*, characteristic while living of deeper zones. This transport of the shells is responsible for rolling and fracture, and indeed it is one of the features of the belt of variables that there is a very large proportion of broken tests, as compared with those of other belts, for fracture is not likely to occur to a great extent where the organisms are embedded in deposits formed on a floor where the water is tranquil. As a caution, however, it may be noted that carnivorous animals, such as crabs, break up the shells of organisms on which they feed, and fracture is so produced to some extent even in deep water.

CHAPTER VIII

THE BELT OF VARIABLES (*continued*): DETAILS OF SEDIMENTS

Examples of Variation. Some examples may be given of variations in the nature of the sediments of the belt of variables. Though not a marine tract, reference may first be made to the Lake of Geneva, which illustrates very clearly on a small scale the great lateral variation in the character of the deposits. The sediments of the floor have been described by Miss Coit[1], and her paper may be studied in this connection.

The variations in true marine deposits are well illustrated by those of the Irish Sea. These have been the subject of study by the late Prof. Herdman and others, as before mentioned[2]. In the report of this committee, after noting the different mechanical (terrigenous) deposits of the area, special attention is drawn to calcareous deposits, generally resembling the deep-sea oozes, but differing inasmuch as the component organisms (nullipores, echinoids, mollusca, etc.) are chiefly benthonic and not planktonic. One such deposit is mentioned off the Calf of Man. The occurrence of these calcareous patches due to the distribution of currents which do not carry mechanical

[1] Coit, G. E., "Nouvelles Récherches sur la Sédimentation dans le Lac de Genève," *C. R. du Congrès Intern. de Géogr.* (Cairo, 1925), vol. II, pp. 59–69.

[2] *Op. cit.* on p. 75.

sediment to these spots is a matter of great interest, as showing one way in which organic deposits can be formed in a state of fair purity on the continental shelf.

In the case of such enquiries as that in which we are now engaged, very valuable information is supplied by study of the ancient rocks, and the Silurian rocks of Great Britain afford a good example. There is evidence that a sea tract extended through Wales, the Lake District, and the Southern Uplands of Scotland, with its northerly coast extending in a W.S.W.—E.N.E. direction somewhere in the centre of Scotland, but shifting from time to time owing to epicyclic changes; and its southern coast running on the whole in a S.W.—N.E. direction from the Bristol Channel to the east of the Welsh borders, and continuing from thence under the site of the Pennines. Each of these coasts had a continental shelf in front of it, with deep water beyond in the centre of the area. On the northern continental shelf the Silurian rocks of Girvan were laid down; on the southern one those of the Welsh Borderland. Outside the shelf in Valentian (Llandovery) and Wenlock times mud was deposited continuously in the deeps beyond the shelves, extending to the centre of the water area. On the shelves all sorts of sediment were deposited; pebble-beds, sandstones and grits, silts, fine muds and limestones, some, like the *Phacops elegans* bed already mentioned, probably due to winnowing action; others, like part of the Wenlock Limestone, largely to direct accumulation of calcareous material where currents carried little mechanical sediment, or, if carrying it, took it right over the spot without depositing it there. Much of the Wenlock Limestone is in fact of the nature of a coral reef. There is much evidence that these Silurian rocks of the shelf were largely due to intermittent deposition and winnowing, as shown by their slight thickness as compared with the deposits of the same age in North Wales and the Lake District. It is very difficult to estimate average thicknesses in the case of shelf sediments of this kind, but they are clearly very much thinner than the corresponding sediments deposited by the "embankment" method on the borders of the deeps, as in North Wales and the Lake District. Owing to this method of deposit, there is usually in the case of the British Silurian rocks a broad zone of transition between the normal sediments of the belt of variables and the fine deposits of the mud belt, marked by coarse sediments of the "embankment" type, such as the Gala beds of the Moffat and Rhayader areas and the Bannisdale Slates of the Lake District. But in the case of the Wenlock beds and to a large extent those of Birkhill age this transitional stage occurs less frequently than at other times, and the mud belt extends to the edge of the shelf.

The nature of these "embankment" deposits is worthy of further consideration, and is well shown by study of the Lower Ludlow rocks of the Lake District. Owing to rapid epicyclic changes they differ in degree of coarseness in vertical sequence, the coarsest, of which the Coniston Grits form an example, being greywacke grits of which the particles are sometimes large enough to constitute a fine gravel. The finest, the Upper Coldwell beds, are laminated silts, which would form an inner edge to the true mud belt. Midway between these in character are the Bannisdale Slates, which are banded beds with alternating layers of silt and very fine grit. These banded sediments require further notice. They are "varved" beds, such as are found in deposits of other ages in various parts of the world, and have certainly been formed in more than one way. The Bannisdale Slates may be taken as an example of one type. Though spoken of as varved beds they are not true varve clays, inasmuch as their formation by glacial action seems to be here out of the question, for the reason, among others, that they are entirely devoid of boulders. Such Silurian rocks of Britain were long ago regarded by C. Lapworth as having been formed under the same conditions as the Flysch of the Alps and adjoining regions. The present writer has suggested that such varved clays may give some clue to the rate of accumulation of sediments[1], and took the case of these Bannisdale Slates for the purpose. They have a thickness of 5000 feet. Measuring the number of stripes of alternating fine and coarse sediment, which are regarded as having been formed by annual accumulation, a coarse and a fine stripe each year, it was calculated that a period of approximately 700,000 years would be required for the deposition of the whole. There are two graptolite zones in the deposit, each therefore suggesting a period

[1] Marr, J. E., "A Possible Chronometric Scale for the Graptolite-bearing strata," *Palæobiologica*, vol. I, 1928, p. 161.

of 350,000 years. Taking the number of recorded
graptolitic zones in the Ordovician and Silurian respectively as 16 and 21, and assuming that each
occupied the same number of years for its formation
as those of the Bannisdales, which is admittedly a
very large assumption, it would need about 13,000,000
years for the deposit of the two systems. No stress is
laid on these figures, but it is suggested that extended
use of this method may lead to some approximation
to the order of magnitude of the geological periods.
It may be noted that other Silurian rocks of the
district are also varved, to wit, the Wenlock muds
and the Lower Coldwell beds, but the banding is very
much finer than that of the Bannisdales, for these
muds and fine silts were obviously deposited much
more slowly.

Deposits of the Continental Shelf of the Open Sea.
Considerable complications are introduced by the
occurrence of confined water tracts, such as bays,
straits and gulfs, as well as deltaic outgrowths in the
tracts where the belt of variables exists. The last-
named have already been touched upon, but will
require fuller treatment in the sequel. It will be
convenient to begin with the deposits of the continental shelf as generally developed in open waters,
omitting for the present these special cases. For our
purpose the classification of Sir John Murray will be
adopted, with some slight modifications. He divided
the deposits into terrigenous and pelagic, the former,
laid down near the land, being formed of mechanically
deposited material, the latter, of the nature of oozes

in the more open sea. But, as shown in the report on the Irish Sea already quoted, many organic deposits are formed mixed with terrigenous ones in the belt of variables. Such deposits have been spoken of as neritic, but the term *benthonic* is more convenient and will be adopted here. We shall consider three main classes of deposit: the terrigenous formed by mechanical deposition, the benthonic essentially by bottom-living organisms, and the planktonic by free-floating or swimming organisms. A fourth division, that of the chemical deposits, is perhaps required, though such deposits are rare in the open ocean. The sediments of the belt of variables are essentially terrigenous and benthonic.

Sir John Murray divided the terrigenous deposits into sands and muds, separating each class into varieties: the sands into ordinary terrigenous (with varieties red and green), volcanic and coral. The muds are similarly divided, an alternative name for the ordinary terrigenous type being blue muds. But in addition it is important to consider the pebble-beds separately, and a beginning of the account of the sediments of the belt of variables is here made with the conglomerates.

Conglomerates. At the present day uncompacted conglomerates occur as beaches, but all conglomerates are not beaches in the true sense of the term. It is only in exceptional circumstances that a true beach can be permanently preserved. With a sea-floor emerging to form land, the beaches thus raised are destroyed by subaerial erosion, before the next submergence; as

each beach is raised a new one is formed at the sea
margin, to be itself destroyed as the emergence continues. With a subsiding area and encroaching sea,
the beach material is redistributed by currents, the
smaller pebbles being carried seaward and deposited
on the foreshore or even farther out, as more or less
isolated patches or individual pebbles, reposing on the
sand, as may often be witnessed on modern coasts
at low water. The remaining pebbles of the beach are
meanwhile reduced in size by rolling until they become
small enough to join the others, and thus the true
beach may be totally destroyed. Currents of exceptional strength may remove considerable masses of
the sand among which the rearranged pebbles lie, and,
as the result of this winnowing action, pebble-beds
with little fine matrix may be produced, just as are
the gravel layers of desert regions. In this way alternations of sand with isolated pebbles, and of actual
pebble-beds, may be produced, and such are often
found among the ancient sediments. These are not
true beaches. It may be remarked that the transport
of pebbles by currents too feeble to remove the much
finer sand in which they are deposited, in the case
where the pebbles are sporadically distributed through
the sand, is due to their roundness, which permits of
their being rolled on the floor by currents not sufficiently powerful to carry them in suspension.

The coarse deposits formed in the open sea usually
consist of well-rounded pebbles, the angularities being
removed by constant rolling. They are therefore true
conglomerates. But occasionally the coarse sediments

are formed under circumstances which prevent rounding of the fragments. Such conditions, as will be seen later, are more often met with in gulfs, fjords and lagoons. If rounding does not take place in any marked degree, breccias are formed in place of conglomerates. Sometimes, even facing the open sea, the land ends in steep cliffs, which plunge down under the water and are succeeded by subaqueous slopes of considerable steepness. If this occurs, fragments detached from the cliffs may give rise to submarine screes, extending into deep water, and even great masses of rock may be detached from the cliffs, sliding or rolling down into deep water, thus escaping attrition and remaining angular, as also do the smaller fragments with which they are associated, giving rise to breccias. The detachment of large fragments, often very large, is facilitated if the land is being pushed seaward along a thrust plane cropping out below the water. This process is exemplified in the case of some of the conglomerates and brecciated rocks among the Triassic strata of the Alps. If ultimately overridden by the rocks above the thrust, these true sedimentary deposits might be mistaken for fault breccias. The steep slopes often found at the margins of coral reefs, like those below cliffs above described, are also favourable for the formation of breccias rather than conglomerates.

The composition of the conglomerates may now be considered, taking first the pebbles and subsequently the matrices. Conglomerates have been described as monogenetic and polygenetic, according as the pebbles are formed of one kind of rock only, or of

many kinds, though even in a monogenetic conglomerate one may expect to find a few pebbles of a
kind differing from those which are dominant.

These monogenetic conglomerates are characteristic of tracts where one kind of rock only forms the
coast: they are well exemplified on the British coast-
lines composed of Upper Chalk. There the pebble-beds
are almost entirely formed of flints, although an
occasional Chalk pebble has escaped destruction. This
also illustrates the fact that pebbles are mainly composed of the more resistant rocks, as the less resistant
are disintegrated into finer fragments which go to
form sands and muds.

Monogenetic conglomerates are found in volcanic
and coral islands, the pebbles of these being formed
respectively of volcanic material and coral limestone.

Polygenetic conglomerates are found where rocks of
different kinds form the neighbouring coast, especially
where along-shore currents transport the material.
They often contain a very great variety of pebbles
when the coast-line is formed of glacial morainic
material. The pebbles of the glacial deposits have
themselves frequently been transported for long distances, and accordingly conditions are specially favourable to the incorporation of a variety of pebbles in a
conglomerate due to their erosion. It is due to this
cause that the pebble-beds over the greater part of the
British coasts are largely polygenetic.

The matrix in which the pebbles of a conglomerate
are included is usually sand or gravel. A word must
be said with regard to the inclusion of the finer

material among the pebbles. An advancing wave may bring on to the beach fragments of varying degrees of coarseness from pebbles to sand. The retreating wave is not powerful enough to drag back all of the pebbles, some being left stranded on the beach. In ordinary circumstances the finer material would be carried back, but a large part of the water of the wave sinks into the interstices between the pebbles, carrying with it much of the finer material, and as the flow of the retreating waters thus imprisoned is checked by friction a considerable part of the finer material is entrapped in the interstices.

Though sand is the dominant matrix of the conglomerates, mud and limestone are by no means rare. Pebbles may be rolled down into a deposit of argillaceous or calcareous mud in the case of steep slopes, such as were above described, or should a tract of argillaceous or calcareous mud lie just beyond the beach, instead of the more usual sand, mud will be introduced into the pebble-bed instead of sand. In the case of the erosion of coral reefs, formed entirely of limestone, the matrix of the pebble-beds is perforce calcareous.

Where desert conditions affect the land adjoining a coast, and river action is in abeyance, the amount of mechanical sediment borne to the sea may be small and even negligible. The mechanical belt may be narrowed or even disappear entirely. As a result of this also the sandy matrix of a conglomerate may be replaced by one of mud or even carbonate of lime.

Sands and Sandstones. Murray's classification of the sediments has already been given: it is desirable at this point to comment upon his use of the term mud, as bearing upon the difference of composition of the sands and muds. He adopts the term mud instead of clay because clay in a pure state is aluminium silicate, whereas the muds consist largely of small chips of various minerals mixed with a varying amount of amorphous matter. Many of the mineral chips in the muds are of the same kind as those composing the sands but of smaller size. In fact to a great extent the sands differ from the muds in size of grain rather than in chemical composition. Nevertheless there is a difference, due to the fact that the less resistant minerals are broken into smaller pieces than the more resistant, and accordingly the highly resistant quartz is found to contribute largely to the formation of the non-calcareous sands, while silicates, especially aluminium silicates, will occur in greater quantity in the muds. Thus quartz occurs largely in the sands, often forming the greater part, sometimes nearly the whole of the deposit, and quartz sands are the dominant deposit of this group, though flint and chert may give rise to sands in a place where these rocks are undergoing erosion.

Small chips of other minerals are usually associated with the quartz, including felspar and various ferromagnesian minerals, as mica, augite and hornblende, while smaller grains of "heavy minerals," such as magnetite, ilmenite, zircon and garnet, occur in most sands. The proportion of the minerals other than quartz depends on various circumstances, especially the nature of the rocks eroded to supply the sediment, and climatic conditions. Mica often occurs in such abundance as to form micaceous sandstones, the large flakes of mica lying with their cleavage surfaces parallel with the planes of stratification. On account of their flatness mica flakes can be carried by feeble currents to a much greater distance than the thicker particles of other minerals of the same weight, and accordingly we find very large flakes of mica abundant in the sands, while the smaller flakes are carried farther out to help in the formation of micaceous muds. Flakes of mica are often found crowding the surfaces of planes of lamination, to the formation of which they undoubtedly contribute, and in some cases at any rate are probably solely responsible for their production. The whole question of the causes of lamination requires fuller study.

In addition to the non-calcareous components of sands, there is usually a certain amount of calcium carbonate, chiefly due to the inclusion of organic remains, the larger of which may be regarded as extraneous matter, but small organisms, both calcareous and siliceous, namely foraminifera and radiolaria, are rarely entirely absent and often contribute to the matrix of the sand. The amount of carbonate of lime varies, and when it is over 50 per cent. the deposit may be regarded rather as an arenaceous limestone.

A certain amount of iron compounds is probably originally present in most sands, though conditions

M 8

are more favourable to the introduction of a large quantity in areas other than the open sea of the variable belt. Whatever the state of the ferruginous material when deposited, it usually changes ultimately to some form of hydrated oxide. Some is formed as carbonate. The frequent occurrence of weathered calcareous sandstones of a deep brown colour, whereas the unweathered rock beneath is blue, indicates the presence of this mineral, though in some cases the iron carbonate may have been introduced into the rock by metasomatic action, replacing lime carbonate. Iron sulphide is also often developed in sandstones as a result of reactions in the presence of decomposing organic matter, both animal and vegetable, though it is also found in other rocks besides sands, including those belonging to other belts than that of variables.

The size and shape of the constituent grains of the sands require some notice. The term grit is often used in contradistinction to sandstone: it is not clear wherein the difference lies. Some writers speak of grits as composed of larger grains than sands, while others apply the term to rocks having grains of greater angularity. Others again confine the term to rocks in which the component particles vary much in size, this being responsible for the retention of a rough surface when broken. It is doubtful whether it is desirable to make any distinction.

The component grains of the marine sands are largely subangular rather than spherical, thus differing from the sands which have been affected by long-continued wind action in desert regions. Being lighter in water than in air, they are held in suspension when moved by currents to a greater extent than when carried by wind, and therefore are not rubbed against the sea-floor for so comparatively long a time during their transport as are the wind-borne grains upon the desert-floor. Furthermore the water acts as a lubricant. The original shape of the grains is determined by mechanical fracture during weathering and they therefore tend at first to be angular, and for the reasons above stated this angularity is not modified to any great extent by subsequent sea-transport. The angles become blunted, but not destroyed, and the resultant form is usually far from spherical. There are circumstances, however, in which the spherical form is closely approached or even arrived at. Mr A. R. Hunt[1] described a case in the English Channel where such a sand is being accumulated. Here the currents are at present not sufficiently powerful to transport the grains elsewhere, or even to raise them above the floor to any great extent. When agitated by the currents, they are rolled to and fro on one spot, and they are not covered up by fresh material. As this process may go on for a long time without change of conditions, very complete rounding results.

It should be noticed that a marine sand may be composed of spherical grains which had their shape determined at a period long before the actual

[1] Hunt, A. R., "The Evidence of the Skerries Shoal on the Wearing of Fine Sands by Waves," *Trans. Devonshire Assoc. Adv. Sci. Lit. and Arts,* vol. XIX, 1887, pp. 498–515.

deposition of the sand. Thus along parts of the English coast where Triassic sandstones of desert type form the coast-line, derived Triassic grains will form the bulk of the modern sand, and it might at first be inferred that the character of the grains was due to marine erosion at the present day.

There is often a considerable amount of argillaceous matter intimately mixed with the sand-grains of the arenaceous sediments. In ordinary circumstances, being of finer grain than the sand, it would be carried farther away, but mud has the peculiar property of cohering into granular masses when brought into contact with salt water, and such particles tend also to coalesce around sand-grains. This was, I believe, first pointed out by Dr Sorby, and it appears to be a frequent cause of the occurrence of a varying percentage of mud mingled with the sand.

Having dealt with the normal terrigenous sands formed in shallow waters around the greater part of the continental areas, we may now proceed to consider the variants, which are only formed in exceptional circumstances and therefore have a limited distribution both in space and time.

1 a. *Green sands.* The term greensand is used for deposits containing glauconite, which gives the rock a green colour. Glauconite is a crypto-crystalline mineral, composed of silicates of alumina, potash and iron, whose exact composition has not been ascertained. It occurs in irregular grains of small size, averaging that of a large pin's head. It is sometimes found filling the interior of organic tests, especially

foraminifera, though it also spreads out beyond the boundary of the tests. The green sands of Murray are arenaceous deposits containing glauconite grains, and he distinguishes those deposits with a finer matrix as green muds; geologists however are prone to call all glauconitic sediments greensands. But it is only with the deposits having a sandy matrix that we are here concerned. Apart from the glauconite grains, the greensands possess the normal matrices of the typical arenaceous, argillaceous or calcareous sediments. We have therefore to consider only the conditions giving rise to glauconite, which must be to some extent different from those controlling the formation of ordinary terrigenous and calcareous deposits.

Glauconite deposits are being formed at the present day in widely-scattered areas, of which the best known is the Agulhas Bank off the southern point of Africa; other areas are in the neighbourhood of the Azores and in parts of the Pacific. They occur in fairly shallow water or on the slopes passing down from the continental shelf, and are said to be produced where there is a meeting of warm and cold currents. Murray regards them as being formed where the sea-waters have a reducing action.

It is generally agreed that the formation of glauconite is due to chemical action in the presence of decomposing organic matter, as suggested long ago by Prof. Sollas[1], who also notices their formation by processes similar to those giving rise to certain nodules

[1] Sollas, W. J., "On the Glauconitic Granules of the Cambridge Greensand," *Geol. Mag.* 1876, pp. 539–544.

of flint and phosphate. Bacterial action probably plays a part in the process. Glauconitic deposits frequently contain phosphatic nodules associated with the grains: examples of this association are supplied by many of the Cretaceous deposits of N.W. Europe, and by certain early Ordovician sediments of Sweden and Esthonia. Judging by their ancient and modern distribution, greensands seem to be found chiefly in the belt of variables, but to some extent also in the mud belt and even in the belt of organisms, where calcareous deposits are found at no great distance from the coast.

Confining our attention to the deposits with an arenaceous matrix, these belong to the belt of variables, and are found in beds of various ages from Cambrian times to the present day. They occur in the Cambrian and Ordovician quartzites of Shropshire, and are abundant in many Cretaceous beds, as notably in the Lower Greensand. In some cases the distribution of the grains is sporadic, not even giving a definite colouration to the rock: in others the amount of glauconite is large, forming a considerable percentage of the whole. In the latter case there is little doubt that the glauconite has frequently been concentrated by erosion of earlier deposits, with removal of the finer material, leaving the coarser grains behind. Such an example is no doubt seen in the case of the Cambridge Greensand at the base of the Cenomanian, where the grains and the accompanying phosphatic nodules have to a large extent at any rate been derived from the erosion of the Upper Gault. The Cambridge Greensand is really a case of a glauconitic calcareous mud, and not of a true sandstone, but no doubt similar examples occur in arenaceous greensands. In other cases concentration is' the result of penecontemporaneous deposit and erosion with the resulting winnowing action already described in the case of other deposits.

It is probable that the rocks with sporadic distribution of grains are commoner than those with a larger percentage of glauconite. They may escape notice, partly from the small number of the included grains, partly owing to subsequent decomposition of the mineral, such as is known to occur. This might easily cause the destruction of all the grains, where sparsely scattered, while affecting only a small proportion of those of a rock with large content of glauconite. Accordingly the special conditions governing the formation of glauconite may be far more widespread than is generally supposed, and the presence of shoal water, sometimes regarded as one of the essentials for its formation, may only be necessary for its concentration.

The reasons for the restricted distribution of important glauconitic deposits both in space and time are therefore not yet fully worked out. Their distribution in space has already been considered: their distribution in time is similar, inasmuch as here also they appear to be discontinuous, with long intervals of absence. Important deposits are well developed in the Cambrian and Lower Ordovician and not again until Cretaceous times in N.W. Europe, but there are many deposits containing sporadic glauconite grains in various rocks of intermediate date and in others later than the Cretaceous. It is interesting to notice the great similarity between the glauconitic and associated sediments of Lower Palæozoic and Cretaceous times, notably the presence in both of phosphatic nodules and of certain red rocks to be ultimately considered.

In the deposits to which reference has been made, the grains of glauconite occur in a form usually described as concretionary, though they are not concretions in the strict sense of the term. In addition to these, films of glauconite have been recorded formed round sand-grains. Similar films but composed of a chloritic mineral are often found encrusting such grains, as in the case of the "greywackes" of our Palæozoic rocks, and these may have some affinity with the glauconitic films as regards origin.

Geologists may be able to obtain evidence sufficient to settle the question whether the meeting of cold and warm currents plays a part in the formation of greensands. One piece of evidence may be mentioned. It will subsequently be noted that E. B. Bailey argues for the formation of the Chalk of England under conditions sufficiently hot to cause a desert area on the adjacent mainland. In such circumstances cold currents are not likely to have existed in the Chalk sea of that region.

1 b. *Red Sands.* Little need be said about the red sands, which have been a matter of interest since the days when Sir Andrew Ramsay gave his well-known

Presidential Address to the Geological Society in 1871. They are connected with certain red muds which will be mentioned later and are formed under exceptional conditions, chiefly, so far as the aquatic tracts are concerned, in restricted waters, the deposits of which will be described later. At the present time they are formed, so far as open waters are concerned, off the mouths of large rivers draining land areas composed of igneous and metamorphic rocks, of which the component minerals, when affected by chemical weathering of the lateritic type, furnish the source of supply of the iron oxide that forms the colouring matter of the sands, being deposited as pellicles around the constituent grains. Modern examples are furnished by the sands now being formed off the mouths of some of the Brazilian rivers.

Among the ancient marine sediments sands of this special character compose the greater part of the Caerfai group of the Lower Cambrian rocks of South Wales, and like the modern deposits probably owe their character to the denudation of adjacent masses of igneous and metamorphic rocks of Pre-Cambrian age.

Having considered the variants of the normal terrigenous sands, we may now pass to the volcanic and coral sands.

2. *Volcanic Sands*. While describing the volcanic sands it will be convenient to consider the volcanic muds at the same time, for they only differ in texture, and both may be formed under similar conditions, though the particles are, as in other cases, largely

sorted according to size, the finer ones, forming the mud, being carried farther from the source of supply. The finer particles of volcanic dust emitted during an eruption may be carried for long distances from their source, falling on land or sea alike. Considerable showers of volcanic dust sometimes fall in the north of Scotland after Icelandic eruptions and must there-fore be deposited in still greater quantity in the inter-vening sea. It is generally regarded as a fact that the wonderful sunsets observed in this country and else-where after the great paroxysmal eruption of Kraka-toa in 1883 were due to very fine dust in the upper layers of the atmosphere, some of which was carried all round the world. In such cases the distance of transport is far beyond that attained by ordinary terrigenous sediment. Accordingly volcanic dust may be deposited in any part of the world, whether land or ocean-floor, but in most cases it is negligible as a rock-former on account of its small amount in pro-portion to ordinary terrigenous material. That the dust can be thus widely transported is shown by the formation of the deep-sea mud (red clay) generally considered to be of volcanic origin, which owes its character to the absence of non-volcanic material of terrestrial origin.

A further source of supply of volcanic material transported from its place of origin is furnished by floating pumice, which gradually disintegrates: the fragments, no longer buoyed up by the vesicles, fall to the sea-floor. Prodigious quantities of such floating pumice have been met with far out at sea: cases have even been recorded where the progress of sailing ships has been impeded by this means.

Although, as we have seen, deposits containing volcanic material, usually mixed with a considerable amount of terrigenous matter, may be met with at great distances from centres of eruption, it is around the volcanic islands that we find the purest typical volcanic sands and muds. It may be noticed in passing that these differ from any other sediments in being of pyroclastic and not of epiclastic origin. They tend, like other mechanical sediments, to be deposited on the whole with the mud belt outside the sand belt, but also like them, they alternate somewhat irregularly in the belt of variables. In the case of violent eruptions, or of formation of much pumice, fragments of various sizes may be deposited in the sand, and even to some extent in the mud.

These sands and muds, when fresh, are usually grey or even black in colour. The rarity of quartz in them is noted by Murray. Quartz is common in many volcanic rocks, and is found, often in abundance, in ancient sediments of volcanic origin: it is probable that the cause of the rarity noted above is that the deposits described by Murray, which were collected during the voyage of the *Challenger*, were the result of basic eruptions, the prevailing type in modern times.

The marine volcanic deposits may be purely pyroclastic, being formed of fragments showered upon the sea-floor without subsequent attrition by waves and currents, or they may be subsequently modified by

these agencies, giving rise to true sands in the case
of the coarser accumulations. The former are usually
spoken of as ashes and the latter as ashy grits, though
we meet with every gradation between them, and
the name ashy grit is also applied to a mixture of
volcanic and non-volcanic material. Organisms may
be and are preserved in any of the rocks of this group.

Turning to examples of ancient volcanic sediments,
the Ordovician rocks of Great Britain may be considered. A few words are necessary concerning the
physical geography of the area in Ordovician and
Silurian times. A gulf-like expanse of sea stretched
across part of our island, in a general N.E.—S.W.
direction, its north coast being somewhere across the
southern Highlands, and its southern one near South
Wales and the Welsh Borderland. The shallows along
these shores formed continental shelves in the belt of
variables, while the central part was normally occupied by the mud belt. This normal arrangement is
shown by the Silurian deposits, which are essentially
non-volcanic: in Ordovician times volcanic activity
was rife, to some extent on the coastal platform, though
chiefly in the more central area, normally occupied by
the mud belt. Accordingly we find that the black
graptolitic muds of this central tract are largely replaced by volcanic material, consisting of lavas and
ashes at and near the centres of eruption, but grading
off into volcanic sands and muds farther away from
these centres. It is in these peripheral parts that the
marine volcanic sands are so well exhibited. They
consist of the varieties of ash and ashy grit which have

already been described and are quite comparable with the equivalent modern deposits save for their colour, to which reference will presently be made. The character of the fauna is instructive. The graptolites of the black muds are here replaced by a shelly fauna comparable with that of the sediments of the continental shelf, such as are seen in Shropshire. This difference is no doubt partly due to shoaling of the water as a result of rapid accumulation of the volcanic material, and indeed, although most of the eruptions were here submarine, some may have been subaerial. This type of volcanic sediment is displayed especially in North Wales and Westmorland.

It was stated of the Ordovician volcanic sediments of Britain that in general they resemble those of modern times: this statement applies to the ancient volcanic sediments as a whole; there are, however, exceptions due to subsequent changes in the last-named. One of these is difference of colour, due to infiltration of water, which causes chemical changes, especially in the iron-bearing minerals; these tend to possess a green colouration, in contrast with the grey or black colour of those of modern date.

3. *Coral Sands*. The mechanical deposits formed from coral-reef limestones are not easily treated under one head, and further reference will be made to them later. They are found under two markedly different sets of conditions: firstly, on the continental shelves not far from the land; there reef-limestones and the coral sands and muds derived from them alike belong to the belt of variables: secondly, around reef-islands,

especially atolls, where each isolated reef, with its accompanying sands and muds, may be regarded as constituting an outlier of the belt of variables, surrounded on all sides by deposits of the organic belt. In the latter case the reef itself is the "land" which supplies the sediment for the formation of the sands and muds. In the lagoon, a narrow belt of variables is formed close to the shore, beyond which is a mud belt occupying the central part of the lagoon. Outside the reef the belt of variables is more marked, consisting of breccias and conglomerates formed of coral rock, passing out into coral sands and these into muds, which extend to the deposits of the organic belt forming the sea-floor away from the reef.

Associations of similar deposits are frequently found among the sediments of ancient date, as for example among the Jurassic rocks of Britain. The coral sands and muds differ from the ordinary terrigenous ones in their composition, which is essentially calcareous, though in the case of those formed on the continental shelf varying proportions of terrigenous matter may be also present. There are, however, certain features closely associated with the coral sands and muds, though not necessarily characteristic of them, such as chemical precipitates of calcium carbonate and formation of oolitic grains; these will be noticed in a later chapter.

The Muds.

The various kinds of mud included in Sir J. Murray's classification may all be found in the belt of variables, but as they are mainly found in the mud belt, their full consideration will be deferred until the deposits of that belt are described. In the belt of variables their distribution is sporadic. They are produced in tracts over which the currents are too feeble to bring in sand in quantity, but have sufficient carrying power to transport particles of mud. They are often silts, that is, admixtures of mud and sand, and not pure mud. By increase of the sand content these silts pass by gradations into argillaceous sands. The proportion of sand in the silts usually increases as one approaches the coast-line, but there are exceptions. In many cases, as before mentioned, the belt of variables is confined to the continental shelf, but it occasionally extends beyond the shelf down the slope into the deeps, when silt and even sand are deposited there, though usually the mud belt begins at or near the edge of the shelf, either within or beyond it, there being however a tendency to begin outside. When the belt of variables extends beyond the shelf, silt or even sand is there deposited, and when rapid vertical alternations of these or of silt and mud occur, the Flysch type, already mentioned, is produced. These deposits form a passage between the belts of variables and of mud. Their formation outside the continental shelf depends upon narrowness of the shelf or unusual strength of currents, or a combination of the two. If currents are not sufficiently powerful to transport the sand-grains beyond the shelf, mud particles alone are carried there and mud alone is deposited. The sand of the shelf, however, does not remain unaltered. The grains are gradually reduced in size by attrition, by currents of varying direction carrying the grains hither and thither, and the fine particles so produced are then by degrees carried seaward and deposited beyond the shelf in their final resting place as mud. Thus the continental shelf is an area where the material may remain for long in the same quantity, but it is not always the same material: as the older becomes comminuted and removed, fresh coarser material is supplied to take its place, and so the process goes on. The process may be compared with that involved in the formation of soil.

As the result of this winnowing action on the shelf, lamination planes after their formation tend to be destroyed, as the result of the constant re-sorting effect of the conflicting currents. In the depths, however, where the deposits are formed in stiller water, these planes once formed remain permanently. Hence the Flysch type of sediment appears to be confined on a large scale to these depths, and as a matter of fact it often extends into the true mud belt, as evidenced by some of the finely banded sediments of the Wenlock Shale division in Britain. When thus extending, the laminae are naturally much thinner. The coarser deposits of the Flysch type seem thus to be characteristic of steep slopes off the continental shelf, where material coarser than mud is carried over intermittently.

This type of deposit is not always formed continuously. In the Bannisdale Slates of Westmorland the striped silts and muds are frequently interrupted by masses of fine grit, often of some thickness. The physical changes determining the variation from fine to coarse are apparently not very great. Nevertheless the grits are devoid of striped structure. This is probably due to the absence of a property possessed by mud particles, already referred to, namely, that they have the power of cohering one to another, and a lamina, whether of pure mud or of silt, is therefore coherent immediately after its deposition, whereas the particles of pure sand are not in the same way coherent, and when deposition recommences after a pause there is intermingling of the particles laid down before and after that pause; hence in the grits there is neither lamination nor alternation of coarser and finer stripes.

Chemical Precipitates

At one time it was considered that save under very exceptional circumstances and over limited areas chemical precipitates were not formed in the open ocean, but were mainly confined to inland water tracts in arid regions, which had no outlets to the sea, and which were subject to excessive evaporation. It has, however, become manifest in recent years that they can be formed under certain conditions in the open ocean. These deposits consist chiefly of calcium carbonate, but compounds of iron may be formed to some extent: the presence of decomposing organic matter is probably in all cases necessary to bring about their formation. Iron sulphide has been actually seen in process of formation within the tissues of decaying vegetation, and much of the iron sulphide

M

found as a rock constituent is probably formed contemporaneously or penecontemporaneously with the rest of the sediment in which it occurs. Deposits of iron other than sulphide, such as many of the oolitic iron ores, were also probably laid down directly, and are not the result of metasomatic replacement of limestone, as was formerly supposed.

The calcareous deposits can be formed in the open sea, largely in the belt of variables, but they are not confined to it, and may be formed even in the organic belt under favourable conditions; it is in certain partly land-locked areas that they attain their most important development, and the whole matter therefore will be considered in the next chapter.

Organic Deposits

These deposits are found in a state of purity in the belt of variables in tracts where the currents do not carry any terrigenous sediments into the area where they are being formed. Usually a certain amount of fine mud is introduced and the deposit is therefore not purely organic.

These organic deposits may be either siliceous or calcareous, though the latter are far more frequent. It will be convenient to consider the modern deposits first, and then to make some observations upon those of ancient date.

The siliceous deposits are formed of sponges, diatoms and radiolaria. At the present day the two latter are well known as forming abyssal deposits, but there is no reason why under certain circumstances they should not be found in the shallows, as they certainly were in the past. The subject will be discussed later.

Calcareous deposits may be formed of various organisms, plant or animal, such as calcareous algae, foraminifera, echinoderms, and molluscs, but the corals are among the most important. Mention has already been made of patches of calcareous deposit now being formed in the Irish Sea, consisting of calcareous algae, echinoderms, and molluscs. The pure coral limestones devoid of any appreciable amount of terrigenous matter occur in the open sea tracts, but as reefs are also found not far from land on the continental shelf, a certain amount of terrigenous matter is here introduced.

It is now generally known that although we speak of coral reefs, the corals themselves are by no means the sole or in many cases even the major contributors to the formation of the deposit. An examination of the atoll of Funafuti in the Indian Ocean showed that the reef-forming organisms occurred in the following order of importance, beginning with the most abundant: *Lithothamnion*, *Halimeda* (both calcareous algae), Foraminifera, Corals, Hydrocorallines. Other organisms such as echinoderms and mollusca occur far less frequently.

It is sometimes stated that reef-building corals only occur in clear water, and that mud is fatal to them. They certainly flourish more in clear seas, but some at any rate can resist a certain amount of mud, and, as already mentioned, they are found on the continental

shelf. Examination of certain ancient reefs shows
that they were then able to resist a considerable
amount of mud, for in some of the higher deposits of
the Lower Carboniferous rocks of Britain we find
such corals as *Syringopora* building definite reefs in a
deposit of calcareous mud, which contains a considerable amount of terrigenous matter. It is not my
intention to enter into a discussion of the vexed and
very complex question of the origin of atolls and
barrier reefs, though it will be mentioned in a later
chapter.

The existence of coral reefs in the belt of variables
is well illustrated by a case occurring among the
ancient rocks, which is worthy of notice. It has already
been stated that the evidence points to the Silurian
rocks of Britain having been formed in a gulf-like
prolongation of the sea, extending across Britain in
a S.W.—N.E. direction; on the continental shelves of
this gulf variable deposits were formed, extending at
times into the depths, which were however chiefly
occupied by mud. We have here a definite case of an
extensive tract of variables succeeded farther out by
nearly uniform deposits of the mud belt. Among the
variable deposits formed on the continental shelf of the
southern shore of the gulf is the Wenlock Limestone,
much of which is essentially a reef-limestone, built up
of corals like *Favosites*, *Halysites* and *Heliolites*, often
in the position of growth. Organisms other than
corals do, it is true, contribute largely to the limestone,
but this, as has been mentioned, is also the case with
modern reefs. The Wenlock Limestone reefs are also

mixed with a certain amount of terrigenous mud, and are quite comparable with modern reefs. They furnish a definite instance of coral reefs formed inside the inner mud line.

This discussion of coral reefs has led us from the modern to the ancient sediments: to the latter we may now definitely turn. A comparison of the ancient and modern deposits of the belt of variables shows that they were formed under generally similar conditions, the main difference being in the proportion of various groups of organisms as rock-formers, for many of those of ancient date have now become rare or extinct. Calcareous algae and foraminifera have been rock-formers from times at any rate as far back as the Ordovician, or probably earlier, and are still so. In early times we get deposits formed extensively of crinoids, trilobites and brachiopods. The trilobites have long been extinct, while the crinoids and brachiopods have become too scarce to be of importance in this connection: the rôle formerly played by the crinoids has been taken up by the echinids, and that of the brachiopods by the molluscs.

The organisms found in the belt of variables are marked by certain characteristics, by means of which it is occasionally possible to refer ancient deposits to this belt, apart from any study of their lithology, but, as the evidence furnished by the latter is usually conclusive, we are not greatly dependent upon the included organisms.

Land-plants are often drifted out to sea in some abundance, and relics of them fall to the sea-floor, mostly in the tracts near the coast, probably within the belt of variables, and they can be but rarely preserved in quantity in the deposits of the mud belt: it follows, therefore, that when found in a mudstone, interstratified with coarser deposits, their occurrence therein implies that there was no extensive change in the area from the belt of variables to the mud belt and back again, but that the mud deposit is but a local accumulation in the belt of variables. What has just been said about land-plants applies also to terrestrial and freshwater animals, which are, however, less likely to be abundant.

The marine organisms of the belt are marked by great variety and benthonic forms are dominant among the larger organisms, though as already stated a certain number of planktonic forms are carried in at times from the open sea, in addition to some living in the shallows. Planktonic micro-organisms are often found in some abundance, especially foraminifera, but also radiolaria. There is no reason why fairly pure deposits of either of these should not be in process of formation at the present time in the belt of variables: this is a matter which will be more fully discussed in the sequel. The great variety of benthonic forms and the abundance of individuals, which are often marked features, are no doubt due to shallow-water conditions. These allow of efficient aeration of the water, and also favour an adequate supply of food. The numerous organisms, after death, furnish a large supply of animal matter, which favours the abundance of carnivorous gastropods, often a marked feature in the shallows, where the variable sediments are mainly accumulated. Sessile benthos attached to rocks, such as barnacles and limpets, as well as boring forms, such as *Pholas* and *Lithodomus*, are most abundant in these shallows, for the conditions favourable to the exposure of bare rock in the depths must be rare.

In the more recent deposits, from early Tertiary onwards, it is usually easy to gain some idea of the depth at which the deposits were formed by study of the included organisms, often identical with living species, but in the case of earlier sediments the difficulties increase, though even there we get some clue, as for instance from abundance of carnivorous gastropods, and of reef-building corals, if we adopt the course above advocated of including all the coral reefs in the belt of variables.

CHAPTER IX

THE BELT OF VARIABLES (*continued*)

SPECIAL AREAS.

OFF coast-lines which are not marked by important variations from general straightness the ordinary deposits of the belt of variables described in the last chapter are accumulated. Certain variations from simplicity are responsible for departures from the normal types of sediment as there produced. These departures from normality are of great importance to the geologist, and in the study of these he can afford much help to the oceanographer and vice versa.

The principal variations in the sediments are due to the existence of more or less land-locked tracts of water, varying from the moderately open waters of wide bays to those which are almost completely enclosed in fjords and reef-lagoons. They vary greatly in size; some fjord-like inlets are quite small, and from these we pass by gradation to large tracts like the Baltic, Mediterranean and Red Seas.

In addition to these partly land-locked arms of the main oceans, there is one other variant from the normal, where deposits are built out as deltaic growths to form new land. The deposits of this character are of much importance to the geologist and will be considered first.

1. *Deltaic Growths*. These deltaic growths may be either single or coalescent. In ordinary circumstances

they tend to shift the belt of variables by converting its inner part into land and causing the outer part to encroach upon the mud belt. In the case of delta growth the deposits from the outset tend to assume the embankment structure as described in the last chapter. These deposits will at first be built on the normal ones of the continental shelf, but if the growth proceeds to a sufficient extent the materials will be shot into the deeps beyond the shelf, in the manner which has also been previously described. As the coarse deltaic matter encroaches on the pre-existing mud belt, the finer will be deposited farther away from the former coast-line, to contribute to the ever-shifting new mud belt.

The case of greatest interest to the geologist is when the delta is growing in a tract of subsiding sea-floor, with the rate of subsidence varying, and the movement sometimes interrupted by actual pauses. Here the periods of subsidence will be marked by sediments of varying character, according as subsidence or deposit is dominant. During periods of pause the shallow waters will be completely silted up, and a land tract formed, which may last for some time and on which terrestrial accumulations may be formed. In such cases very variable deposits may be met with in a vertical section, with alternation of terrestrial and aquatic formations. The aquatic sediments will form wedges, at first increasing in thickness seaward and even actually dying out landward, causing there the coalescence of two or more of the terrestrial deposits. The beds of aquatic origin will normally contain

marine fossils, though those of freshwater or terrestrial origin may be introduced in ways previously
described, and estuarine assemblages may occur as
subsidiary features, and even fluviatile remains may
be found unmixed with marine in the braided river
channels of the delta: these, however, are all subsidiary to and confined to delta growth: it must be
clearly understood that the typical delta formations
have nothing to do with estuarine formations, though
often confused with them by writers.

A very full account of the characters and structure
of many deltas is given in the first volume of Sir
Charles Lyell's *Principles of Geology*, to which the
reader is referred for further details.

The type of deposit naturally varies in different
deltas according to the degree of coarseness of the
sediment carried by the rivers at their mouths. In
many the prevailing sediment is silt, which is now
alone carried by the Ganges in the lower stretches of
its course. In other cases sand is carried, and in yet
others gravel and pebbles. Indeed, the sequence often
shows rapid alternations of fine and coarse deposits in
the same delta, as shown in the case of the Ganges, by
artesian borings in Calcutta.

So far we have discussed the true marine deposits
of deltaic origin formed in the subaqueous tracts,
which are alone germane to this chapter, but the
terrestrial deposits of the inner part of the delta,
which is above sea-level, though strictly belonging to
terrestrial deposits, which have been described in a
former chapter, must be considered in connection

with the marine deposits into which they pass laterally.

So long as the rate of sedimentation is equal to or greater than that of subsidence, the terrestrial part of the delta grows outward, and the previously formed marine sediments become covered by others of terrestrial origin. This replacement of marine deposits by those of terrestrial origin will begin at an early stage of the formation of the delta at the apex, and obviously must occur at later and later periods away from the apex. So long as deposit keeps pace with subsidence, the thickness of the marine deposits will go on increasing, and as the fluviatile deposits of the flood plain of the land part of the delta are graded to them, they also will grow thicker by aggradation, so that in course of time great thicknesses of this fluviatile material may be formed without any intercalation of marine deposits, which may accordingly contain relics of freshwater and terrestrial but not of marine organisms. When a pause occurs in subsidence the deposit of mechanical material is stopped in the land parts of the delta, which has been in existence for some time, just as in those parts where the sea tracts have been recently filled up and converted into land, though in the latter case the accumulations formed during a pause will rest on marine sediments, in the former on terrestrial. The deposits formed during the pause are largely cumulose and of vegetable origin, though some may be formed of drifted vegetation of local water tracts. The latter are likely to be formed in greater quantity in those parts of the

delta which have most recently been converted from
sea to land.

An interesting example of alternating mechanical
and cumulose sediments is furnished by the aforementioned artesian wells at Calcutta, where peats
alternate with mechanical sediments. It is important
to note that no trace of marine organisms has been
found in these deposits.

So long as subsidence goes on at a less rate than
deposit the land part of the delta will receive terrestrial accumulations only, gravels, sands and muds
during subsidence, surface soils and peat during pauses.
These terrestrial accumulations will pass laterally into
the contemporaneously formed marine deposits of the
submerged tract of the delta. If at any time the rate
of subsidence should exceed that of deposit, the sea
will encroach on the land, and a marine deposit will
be laid down on the previously formed freshwater ones,
to be covered subsequently by fluviatile material
when the rates are reversed. Recurrence of such conditions will produce alternations of marine and terrestrial deposits in those parts of the delta which
have been alternately above and below sea-level. It
is important to note that the above-described phenomena may be produced as the result of subsidence
alternating with pauses, without the intervention of
any uplift.

The effects of changes in the minor channels and
distributaries of a delta are considerable. At times
new channels are formed and may subsequently be
partly silted up, causing formation of islands. Rafts

of vegetation may often develop in channels, giving rise to local carbonaceous deposits of aqueous origin. False-bedding and "wash-outs" may be frequent, owing to contemporaneous erosion. Should these processes go on for a long period in coalescent deltas, when general subsidence is in progress, massive accumulations of this character may be built up on a large scale. The formation of the Coal-measures and similar strata was probably due to such processes, for the similarity to those of modern deltaic growths of the phenomena there presented is very remarkable. The presence of occasional "marine bands" with marine shells, alternating with those containing presumed freshwater shells like *Naiadites* and its allies, in the Coal-measures is quite in accordance with this view.

It has been argued that the sediments of the Coal-measures were laid down in a freshwater area, but from what has been said, it would appear that the character of the sediments can equally be accounted for by building out of the land portions of deltas, formed by aggradation of freshwater sediments into a salt-water tract. It is probable that such conditions as above described were more widespread in Coal-measure times than they are now, though they obtain over large areas in many parts of the world. Vegetable growth on land surfaces over wide areas, formed in similar manner to the peat beds met with in the well-borings of the Ganges delta, would be responsible for the extensive coal seams of the Coal-measures, or at any rate for some of them. Into the much disputed

question of the origin of coal it is not proposed to
enter. The reader is referred to Dr E. A. N. Arber's
book, *The Natural History of Coal*, for a succinct
account of the matter.

Deposits of Partly Land-locked Seas

It has been already noted that several causes tend
to disturb the regular arrangement of sediments in
parallel belts of varying degrees of coarseness, the
coarsest near the coast and finer ones in succession
away from it. These variations from regularity are
specially prone to occur where there are partially
land-locked tracts of water, ranging from bays to
areas with narrow connections with the open sea. I
have observed a good example at Carnarvon, where
the little river Seiont flows into the Menai Straits
through a tiny estuary less than a mile in length.
The estuary is being silted up with mud, while immediately outside its mouth the Straits are receiving
deposits of nearly pure sand. The mud is brought
down by the river, and the deposits are probably
assisted in their formation by the tendency already
referred to for the flocculation of mud particles when
they reach salt water, while the sand-grains not undergoing this process are carried into the Straits. The
sands of the Straits are, however, presumably, not derived in the main from the Seiont, but are brought
from elsewhere by tidal scour, which carries mud into
the open sea, leaving the sand behind.

Other interesting examples of such variations from

normality are found where shingle spits are built across the mouths of estuaries and bays. Such variations in the nature of sediments inside and outside shingle spits are of much interest to the geologist and deserve notice at some length. Mr J. A. Steers has kindly furnished me with an account of his own researches on shingle spits, mainly in East Anglia, which I append here. It will be noticed that strictly speaking some of these deposits are of terrestrial origin.

The East Anglian coast presents many cases illustrating exceptions to the general sequence of shore deposits. Wherever a shingle bar occurs there is more or less a repetition of the sequence. The north Norfolk coast west of Weybourne is bordered by a very shallow off-shore zone, sand-flats extending many miles seawards. At low water vast flats are exposed off the main beaches. A traverse at right angles to the trend of the coast will usually show the following very generalised arrangement of deposits. (See Fig. 7.)

The relative width of these deposits varies greatly. Shingle bars are common along this coast and are complex in pattern, many subsidiary ridges running landwards from the main ridge. Between the older of these lateral ridges are marshes. The newer ridges, when first formed, enclose flat sandy areas: the silt brought in by the tides is gradually deposited in the slack at high water, and a thin film of mud is laid down. With the gradual incoming and spread of halophytic plants more silt is trapped and the low marsh stage is formed. In the inner parts of these marshes

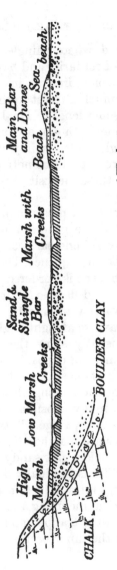

Fig. 7. Generalised section of the Norfolk coast, west of Weybourne.

vegetation spreads more rapidly, more material is trapped, and the level of the marsh rises proportionally. Plants which characterise the upper marshes come in, and typical high marsh is formed. This passes into shingle and sand and then to dunes which nearly always rest on sand or shingle ridges. On the seaward face of the main beaches there is first a sandy and shingly beach and then only sand. In calm weather the separation of the sandy and shingly upper parts of the beach from the sandy lower part is often very sharp. The section shows marsh mud, sand and some shingle enclosed between the mainland and the major shingle bar, but it is not always as simple as this. Irregular patches of shingle and sand often occur, or perhaps odd shingle bars, parallel with the main beach, are found well within the latter. These shingle bars are surrounded by marsh. Again, alongside the old shoreline there may sometimes be seen patches of the former sand and shingle beach, and even if this is now covered for the most part by marsh and vegetation a slight digging usually reveals its presence. Practically the whole area between the mainland and the main shingle bar is much cut up by an intricate series of creeks draining into one or more major channels. In many cases the floors of these creeks are of sand, whereas their sides are of mud, the height of the banks being an index of the amount of mud accretion. But it is worth noting that the amount of sand on the bottoms and sides of the main creeks increases as they are traced toward the open sea. In other cases mud may reappear just seawards of a main shingle bank.

Occasionally this is seen at Orford Ness in Suffolk. In this case sand is practically absent, the whole vast bar being formed entirely of large shingle. But with the gradual landward movement of the bar as a whole, the marsh which it encloses has been over-rolled, and marsh mud thus sometimes shows at the foot of the bar on its seaward side. Another excellent example of this is seen at the south end of Walney Island on the Furness coast of Lancashire, and on one or two occasions mud has been exposed on the sea side of the main bar at Blakeney.

In fact, one of the most striking features of such coasts is the abrupt changes in the deposits. Sand, shingle and mud are found close together. A new shingle bar is produced on a sandy bottom by waves and currents. Within the shelter of this bar marsh mud may gradually accrete. At other times storms may displace masses of shingle and lay them down on older marshes, and dune-sand is often blown over marshes in all states of development.

Some work bearing upon this subject has recently been carried out among the ancient sediments. An interesting example has been described by Dr W. F. Whittard in certain rocks of Valentian age in Shropshire. (See *Quart. Jour. Geol. Soc.* vol. LXXXIII, 1927.)

"Between the villages of Little Stretton and Plowden in South Shropshire, Upper Valentian rocks discordantly rest upon the southern flank of the southwest spur of the Longmynd, which is of pre-Cambrian age. During Valentian times this spur was probably lofty and steep-sided, with southerly flowing streams debouching into the sea-gulf between the spur and the ridge of the Caradoc Range and its south-westerly continuation. The coast-line was rugged and indented, and many alongshore structures were formed which are characteristic of complex coasts, such as shingle accumulations in the form of beaches, bars and shelf-deposits. At the present day many of these structures are again visible, and it is thus possible to define the probable alongshore conditions which obtained here during the Valentian deposition.

"A study of the Valentian sandstones and conglomerates (compacted sands and shingle) shows that they are mainly arranged as sediments filling-up pockets in the coast, as deposits resting upon shelf-like projections from the mainland, or as accumulations upon promontories. Under the first category are included the pebble-beaches: the second is divided into pebble-banks and deposits accumulated round sea-stacks, and the third comprises pebble-bars in various stages of development.

"*Pebble-beaches* (Fig. 8). These beaches, of which there are many excellent examples, are grouped round Hillend Farm and the hamlet of Minton. They may be isolated from the mainland, or connected over an intervening spur, by a thin deposit of shingle. They are small in extent, and the maximum breadth of the outcrop in many cases corresponds with the bottom of a modern valley in the Longmynd. It therefore appears that some of the present brooks flow in valleys which may have been in existence in Valentian times. At Park Plantation, Plowden, conglomerate is now

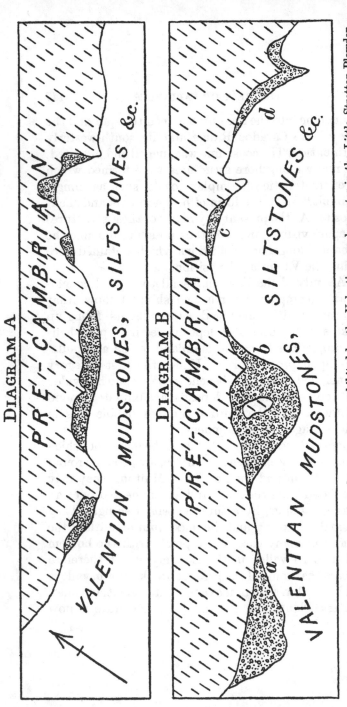

Fig. 8. Diagrams illustrating the alongshore structures exhibited by the Valentian conglomerates in the Little Stretton-Plowden outcrop, Shropshire. *A*. Diagram showing the arrangement of the pebble-beaches round Hillend Farm, Plowden. *B*. Diagram of the shelf-deposits and pebble-bars. (For convenience in drawing and interpretation the examples are not represented in the same sequence as they occur in actual fact.) The direction of alongshore drift is from the south-west to the north-east.

All drawings are on the scale of 6 in. to 1 mile.

found in several small outliers, which are essentially thin veneers of rock occupying small valleys in the Longmynd, trending north-west, south-east.

"*Shelf-deposits* (Fig. 8). Two kinds of shelf-deposits are recognised. In one case sand and shingle apparently accumulated upon a submarine ledge to form a pebble-bank (diagram *A*); in all probability the ledge was regular and gently sloping. The second form of shelf-deposit is characterised by the deposition of shingle round an area of pre-Cambrian rocks, which probably projected through the water as an islet of sea-stack (diagram *B*). At the extreme south-westerly termination of the Longmynd headland near Plowden, where there are two good examples of fossil sea-stacks, quarrying operations have shown that near the coast-line the sea-bed was both irregular and steeply inclined.

"*Pebble-bars* (text-fig. 8, diagram *B*, *c* and *d*). By a study of the alongshore gradation in the Valentian sediments it has been determined that the direction of the transport of detritus along the south-east shore of the Longmynd headland was from the south-west to the north-east. For the formation of pebble-bars it was necessary that locally the mainland protruded out to sea as points or promontories. As the alongshore currents carried detritus, deposition took place in the waters of the embayment to the south-west of the promontory, at first resulting in the formation of a pebble-beach (text-fig., *c*). With further deposition the shingle extended farther along the promontory until finally it projected out to sea as a pebble-bar (text-fig., *d*).

"All the previously mentioned conglomeratic beds are very impersistent and of small areal extent. In the Wenlock Valley, however, the arenaceous beds of Valentian age outcrop over a distance of about 22 miles. It has been shown in a restoration of this particular part of the Valentian shoreline, that shingle was mainly deposited within an open embayment and accumulated to form a beach about 15 miles in length[1]. North-east of this bay, owing to the alongshore drift and softness of the rocks of the immediate mainland, shingle gradually gave place to sand deposits.

"In Shropshire, therefore, we have a fossil coast-line of Valentian age, which is characterised by its attendant sand and shingle accumulations, similar in distribution and structure to deposits now being formed along many rocky coasts."

Attention may next be directed to the effect of land-locked waters on the character of the sediments deposited in them. One important effect of a restricted communication with the open sea is a departure from normal salinity, the land-locked areas being often less salt under cold conditions and more salt in warm latitudes than the open sea. An example of the former is the Baltic and of the latter the Mediterranean and the Red Sea. Variations of salinity may affect both the lithological character of the deposits and their included organisms. In addition to those partly enclosed sea tracts in which the salinity is affected by temperature, there are also estuaries, which may occur

[1] *Q.J.G.S.* vol. LXXXIII, 1928, p. 740.

under any climatic conditions, and these may first be considered.

Estuaries

Certain deposits which were not formed in estuaries are sometimes loosely spoken of by geologists as estuarine. These are often marked by vertical alternation of deposits, some of which contain only marine organisms, and others freshwater and terrestrial forms only. Such deposits may be formed in deltaic accumulations in an area of alternate subsidence and repose, as already explained. These deposits are more accurately described as fluvio-marine than as estuarine. The term estuarine deposits should be restricted to such as are definitely formed in estuaries.

The sediments of this category may vary in a high degree, according to local conditions, such as the size of the estuary, the strength of the currents, the width of the opening, and so on: they need not be discussed at length, for they do not differ in any marked degree from those of open-sea formation, though the deposition of flocculated mud deposits will be more abundant for reasons already mentioned, and pure calcareous deposits are of little importance. The included organisms afford the best test of estuarine conditions, and will be first considered. Attention may first be directed to those actually living in the estuarine waters. An intermingling of fluviatile and marine forms when found is of great importance, but the former need not always be present, especially in the case of large estuaries, where they are likely to be most abundant near the head, where the river enters.

As the waters there will be brackish, some modification in the tests of mollusca, here living under abnormal conditions, may be expected. In addition to
the actual denizens of the tracts, relics of organisms
may be introduced into it in a dead state. Those
brought from the land may first be considered, whether
truly freshwater or terrestrial. As has already been
seen, these may also be deposited in the shallow
waters of the open sea, but they tend to occur far
more frequently in estuarine deposits. Remains of
land-plants are often so abundant as to furnish actual
carbonaceous deposits. Of the animals, both invertebrates and vertebrates occur, the mollusca being most
abundant among the former, and the mammals among
the latter. Relics of dead organisms are also introduced from the open sea with the incoming tides;
these will be largely planktonic, foraminifera and
radiolaria being frequently found, but also relics of
larger organisms, such as crustacea. It follows that
the organisms of the upper waters of the larger estuaries show some resemblance to the freshwater assemblages, while those of the seaward tracts have a close
affinity with those of the waters of the open sea.

An example illustrative of these remarks is furnished
by the fauna and flora of the London Clay, which is
generally regarded as the product of the estuary of a
large Tertiary river. Land-plants occur in considerable abundance in places, having been, as generally
supposed, floated down by the river, sometimes in
tangled masses, as in the case of the Sheppey "Pine
Raft." Remains of land animals are not rare, those

of mammals being most noteworthy. Of aquatic forms we find abundance of marine mollusca and crustacea, and what is more important, of verte-brates more characteristic of rivers and estuaries than of the open sea, such as crocodiles.

Having now considered the special type furnished by the estuaries, some general remarks may here be inserted. In size and general character the partially land-locked sea tracts of the past certainly varied much as they do now, some being small bays, others resembling the Mediterranean. Moreover, then as now those of large size had floors not confined to the belt of variables, but extending into the mud belt and even into the organic belt. In this chapter, however, we are concerned only with those portions included in the belt of variables. Some of the phenomena to be noticed are confined to the restricted areas; others are shared, though many to a minor degree, by the waters of the open ocean, especially in the shallows.

The phenomena displayed in the case of the large areas of restricted sea tracts are of especial importance to the geologist, for such tracts were quite frequent in past times: in fact it is doubtful whether any of the ancient seas of N.W. Europe belonged to the un-restricted open ocean, even such deposits as the Chalk having been most probably formed in large tracts somewhat resembling the modern Caribbean Sea, though these tracts of N.W. Europe certainly in-cluded areas of deposition belonging to the mud and organic belts as well as to the belt of variables.

The Baltic

The Baltic is a good example of a partially enclosed sea of brackish character, owing to its situation in a high latitude and the large amount of river water which is poured into it. It is of interest also because the water becomes increasingly fresh as the north end furthest away from the open sea is approached. The sediments deposited in it can hardly show any important difference from the normal deposits of the open ocean, though there may be variations in detail, such as for instance the supply of carbonaceous matter furnished by aquatic plants, though this varies greatly in the open ocean itself, and no doubt many more variations in the sediments and probably in the distribution of organisms occur in the fjords which indent its western coast.

The distribution of the organisms of the main water tract is more interesting than that of the sediments. These show some features reminiscent of those of an estuary; indeed the conditions are essentially estuarine save in name. The difference between the fauna of the Baltic and that of the open ocean is not marked; indeed, it is so slight that the Baltic is not considered to be a separate marine province, but forms part of the Celtic Province which occupies the seas surrounding the greater part of the British Isles, most of the species being identical. As the result of the exceptional conditions as regards salinity, the fauna is limited, there being few species as compared with the open ocean of the Celtic Province. Certain structural variations occur in some of the species, invertebrate and vertebrate alike. For example, the herring of the north Baltic is not quite similar to that of the south, which resembles that of the open waters. Similar changes are found in some of the mollusca: they are, however, of a trivial character, and such changes in species of past times would probably pass unnoticed by the geologist, save in the case of those of very recent date, such as the shells of the Pleistocene "Littorina Sea." As the northern end of the Baltic is approached the amount of fresh water is so great that freshwater species of mollusca are found living together with marine species; thus *Littorina* is found concurrently with *Limnaea*. This is a matter of importance to the geologist. If the beds now being formed in the Baltic were exposed in continuous horizontal section, we should find at one end deposits with a completely marine fauna, and at the other end a mixture of marine and freshwater forms, thus indicating the direction of the connection with the open ocean. In the case of rocks of ancient date, apart from the occurrence together of freshwater and marine forms, our study would be little facilitated by a knowledge of the conditions now obtaining in the Baltic.

In partially enclosed seas of warm latitudes, on the contrary, much light is thrown on the origin of older deposits by those of similar sea tracts now existing. To a consideration of these we may now turn.

The Rann of Cutch. The Rann of Cutch shows that deposits of salt may be formed not only in salt lakes having no connection with the sea, but also in tracts of water due to temporary admission of the sea to low-lying areas of land. An account of it will be found in Lyell's *Principles of Geology*. The area is somewhat larger than that of Yorkshire and when dry supports a growth of tamarisks and other plants. At times the sea invades it and the water is subsequently evaporated, causing a precipitate of salt. The vegetable growth which previously occupied the land is killed, and its relics may become embedded in any deposits then being formed. As the area is subject to earthquake shocks, subsidence may go on, often rapidly, as in the case of the earthquake of 1819, whose results are described by Lyell, and in such circumstances considerable thicknesses of deposit, partly of mechanical character and partly precipitates, may be formed. Somewhat similar conditions probably occurred on a large scale in past times. Thus the Permian Magnesian Limestone of England and N.W. Europe, although apparently formed in a basin ultimately severed from the open ocean, must have been at the outset connected with it, and the effects of evaporation may well have been appreciable before the final severance. The destruction of tamarisks in the Rann of Cutch may be illustrative of the cause of the entombment of the plants in the Permian Marl Slate and Hilton Shales below the Magnesian Limestone. The chemical precipitates of tracts like the Rann of Cutch are mainly formed after their severance from the sea, but they are so obviously connected with marine sedimentation that their consideration properly falls into this chapter.

The present water tracts of the Aralo-Caspian basin, though now lakes, are relics of a once continuous area of ocean, joined to the Arctic; and, though the present phenomena connected therewith are lacustrine, at an earlier stage they would belong to the category of deposits formed in a restricted tract of sea, though no doubt at first such conditions as suit the precipitation of salts like sodium chloride would be undeveloped.

In a similar way the Magnesian Limestone of Durham, with its scant fauna, appears to have been due to the extension of a gulf from the Permian sea of southern Europe and Asia, in a north-westerly direction, the sea at first not reaching as far as Britain, for the character of the Durham deposits would indicate that that tract was not submerged until the gulf had been severed from the open ocean; otherwise the Durham fossils would be of more open-water character than is actually the case.

The deposits of the Aralo-Caspian basin probably illustrate the formation of the Permian rocks of N.W. Europe in more respects than that just indicated. The salinity of these water sheets differs largely owing to differences in size, the larger being the fresher, their salinity being such as allows of the prevalence of ordinary marine organisms, not greatly differing from those of the open sea, though these become modified as evaporation goes on: this point has been already noticed in Chapter V. In the smaller lakes organisms are rare and finally become extinguished. Concurrently with these changes salinity increases and may proceed to complete desiccation, causing the deposition of highly deliquescent salts. Between these water tracts terrestrial accumulations are being formed: the conditions of the area, therefore, as regards sedimentation are very varied. The nature of the Permian deposits of N.W. Europe seems to point to similar complex conditions having prevailed there in those times.

To return once more to the Rann of Cutch, the conditions there also are as stated very complex. Another complication is introduced by occasional incursions of fresh water from an arm of the Indus, which sometimes discharges its waters into the sunken tract. In this way bands of freshwater deposits containing fluviatile shells may be interstratified with others indicative of highly saline tracts occupying an arid region.

Lagoons of Coral Reefs and partially constricted Sea-water Tracts akin to them

Though the deposition of chemical precipitates on a large scale in the sea was formerly doubted, it is now recognised that it can and does take place in favourable circumstances. It is regarded as possible by some that such precipitation may take place, at any rate to a slight extent, in open waters of great depth. Whether this be so or not, it has been made abundantly clear as the result of recent work that it does occur in more or less confined tracts in the shallows, chiefly in the belt of variables, though in certain circumstances similar precipitation may go on in the organic belt. The barriers need not necessarily rise above water-level. Tracts of water partially confined by shoals rising nearly to the surface may produce the necessary conditions. Such conditions occur in the case of barrier reefs and atolls, where precipitation may go on not only in the lagoon, but also on the submerged shelf outside. We may first consider the deposits connected with barrier reefs and atolls, and afterwards those formed in such shallow waters as are not normal coral reef-lagoons.

Coral Reef-lagoons. Chemical deposit does not seem to occur extensively in the lagoons, and what does take place is no doubt similar to that which is found in the tracts to be presently described; this includes the formation of oolitic grains, which may, however, be mentioned here as frequent accompaniments of coral rocks, ancient and modern. One change, though a metasomatic one, is a very frequent feature in reef formation: this is dolomitisation. The conditions in the lagoon are more favourable to this reaction than those in the waters outside. The high temperature and consequent evaporation will increase the salinity of the water. The magnesium salts of the water will react with the calcium carbonate of the calcareous mud, and as water percolates through the floor to the reef rock below this will also be affected and some of its calcium carbonate replaced by magnesium carbonate.

In connection with this subject some experiments made by Klement[1] are of interest. He submitted

[1] Klement, C., *Bull. Soc. Belge Géol.* vol. IX, 1895, p. 3.

powdered aragonite to the action of magnesium sulphate in a saturated solution of sodium chloride, at a temperature of 91° C. The result was a replacement of some of the lime by magnesia, to a maximum of 42 per cent. The product was a mechanical mixture and not a compound. In lagoons the temperature and the concentration of sodium chloride do not approach those in the experiments, but Klement points out that with lower temperatures and less salt the change does occur, though to a less extent. Skeats suggests that the increased pressure, up to 5 atmospheres in the deeper parts of the lagoons, facilitates the process[1].

The bearing of Klement's experiments on dolomitisation of limestones formed in the highly saline waters of the inland lakes of arid regions is obvious. These waters must attain higher temperatures and greater salinity than those of reef-lagoons.

It is doubtful how far deposits formed in true lagoons of atolls and barrier reefs have been detected among the ancient rocks, though Dupont has claimed their occurrence in the Upper Devonian rocks of the Ardennes, and apparently on good grounds[2]. At any rate the required conditions no doubt obtained in other lagoonal tracts.

Seas off the Bahamas and Florida

In addition to true coral reef-lagoons there are water tracts under somewhat similar conditions which

[1] Skeats, E. W., *Quart. Jour. Geol. Soc.* vol. LXI, 1905, p. 97.
[2] Dupont, E., *Bull. Musée Royal Hist. nat. Belgique*, vol. I, 1882, p. 89.

have been the subject of study in considerable detail in recent years. Prominent among these is the shallow sea off the Bahamas and Florida. Abundant reefs cause constriction of the bottom waters even when the reefs do not actually reach the surface and the waters here being at a high temperature, considerable evaporation takes place. It is here that we meet with most satisfactory evidence of production of deposits of calcium carbonate by precipitation. These deposits are therefore of high importance to the geologist.

Some light has been thrown on the formation of precipitates of $CaCO_3$ in marine tracts, and the formation of oolitic grains therein, by G. H. Drew[1] and T. Wayland Vaughan[2]. The particular deposits described are forming in the bights and channels of the Bahamas and behind the Florida reef: that is in shallow enclosed waters. The conclusions arrived at by Vaughan are stated as follows:

"1. Denitrifying bacteria are very active in the shoal waters of both regions (Florida and Bahamas) and are precipitating enormous quantities of calcium carbonate, which is largely aragonite.

2. This chemically precipitated calcium carbonate may form spherulites or small balls which by accretion may become oolite grains of the usual size, or it

[1] "On the Precipitation of Calcium Carbonate in the Sea by Marine Bacteria," and "On the Action of denitrifying Bacteria in Tropical and Temperate Seas," *Papers from the Marine Biological Laboratory at Tortugas*, vol. v, 1914, No. II.

[2] "Preliminary Remarks on the Geology of the Bahamas with special reference to the origin of the Bahaman and Floridian Oolites," *id*. No. III.

may accumulate around a variety of nuclei to build such grains."

It is however only fair to state that C. B. Lipman[1] disputes the validity of Drew's conclusions as to the effect of bacteria, and offers alternative explanations of the precipitation of $CaCO_3$, either by the alteration of the physical conditions of sea-water or by the abstraction of CO_2 from sea-water by plants.

Similar Conditions indicated by Ancient Rocks

The formation of oolites is a frequent occurrence among the ancient rocks. In Great Britain excellent examples are found among the Lower Carboniferous and Jurassic limestones, sometimes, though not invariably, in connection with coral reefs, but always under conditions which suggest shallow water and high temperature. A further study of these structures would be of interest, in order to ascertain whether their formation occurred concurrently with the precipitation of the calcium carbonate forming the matrix of the rock. That this precipitation will occur without the accompanying formation of oolite seems to be clear, and a case appears to be furnished by much of the material of the English Chalk.

A very remarkable contribution to our knowledge of precipitation of calcium carbonate under conditions somewhat similar to those just described has

[1] "A Critical and Experimental Study of Drew's Bacterial Hypothesis on $CaCO_3$ precipitation in the Sea," *Papers from the Marine Biological Laboratory at Tortugas*, vol. xix, 1924, pp. 179–191.

been made by Mr E. E. L. Dixon[1] in the case of certain Carboniferous deposits of South Wales, belonging to the Lower Carboniferous system and to the *Posidonomya* beds placed by some writers at the base of the Upper Carboniferous. He describes the area in which they were formed as a "lagoon," but explains that he uses the term in a very wide sense, for a shallow-water tract not necessarily constricted by barriers, submerged or otherwise, lying to seaward. This will be referred to later, when it will be argued that a much wider area than South Wales was of a gulf-like character. Mr Dixon describes four types of "lagoon" deposit in South Wales, of which only one concerns us at present, the rest being considered in the following chapter.

The type now to be considered is a very fine limestone, once obviously a calcareous mud. The rock as it now exists has a porcellanous appearance. Some of the belts form pseudo-breccias, due to accumulation followed by conversion to a partly coherent state, after which the sediment was stirred by gentle currents, with resultant fracture and subsequent consolidation. Some of the limestones are pisolitic. Mr Dixon gives good ground for supposing that these fine limestone muds were due to precipitation of calcium carbonate in the shallow water. Their characters strongly recall those of the deposits near the Bahamas and Florida. Judging from the general character of its organisms, the Carboniferous Limestone of Britain

[1] Dixon, E. E. L. and Vaughan, A., *Quart. Journ. Geol. Soc.* vol. LXVII, 1911, p. 477.

was formed under warm climatic conditions, as indicated by the abundance of reef-corals.

Prof. Tarr[1] has advocated a somewhat similar origin for some of the deposits of the Chalk, believing that they also were formed as chemical precipitates, contrary to the view formerly held that they were of organic origin. These deposits also may well have been formed in a constricted water tract, as will be argued later.

Both these and the Welsh Carboniferous deposits just described may well be regarded as belonging to the organic belt, but in each case the mud belt was probably absent, and the organic belt came close to the shore. From the similarity of these deposits to those of the Bahamas and Florida they are treated here under the heading of the belt of variables. As regards the absence of the mud belt, it is probably due during the formation of both the Carboniferous and Cretaceous calcareous muds to arid conditions on the adjacent lands, causing a minimum amount of mechanical sediment to be transported into the sea. This has been ably advocated in the case of the Chalk by Mr E. B. Bailey[2]. It is probable that as the result of future work evidence will be brought to light showing that many other limestone masses among the older rocks were due to chemical precipitation, probably in the presence of decaying organic matter.

Iron compounds. The presence of iron compounds in sea-water is well known, and is important in con-

---

[1] Tarr, W. A., *Geol. Mag.*, vol. LXII, 1925, p. 252.
[2] Bailey, E. B., *id.*, vol. LXI, 1924, p. 102.

nection with the partial or entire formation of certain ferruginous deposits. Special concentration of these iron compounds is likely to occur in constricted tracts, such as are now being described, and there is indeed a considerable amount of evidence that such is the case. Besides mere pellicles formed around grains of other materials, fairly pure deposits of iron carbonate, or hydrate, may be formed.

In addition to this, definite deposits of iron compounds are known to occur as marine formations, some as the result of metasomatic change of lime carbonate but others directly formed as iron compounds. Such are well known in certain lakes, as for instance limonite in certain Swedish lakes, apparently formed in the presence of organic matter. And there seems no reason why they should not also be formed in the sea, and no doubt the conditions obtaining in partly confined water tracts would be most favourable for their formation.

Deposition of the Old Red Sandstone

I may conclude this chapter with some comments upon the deposition of the Old Red Sandstone, as an illustration of the difficulty of ascertaining the conditions of deposition of strata in which the organisms are of extinct species. Many of the beds of this formation are also coloured red by the pellicles of iron oxide above alluded to.

It is well known that various views have been held as to the mode of formation of these deposits, nor are writers by any means agreed even now. They have been regarded as fluviatile, lacustrine, brackish water, and marine. It is probable that no one explanation is applicable to the whole of the deposits, for there is much variety both in lithology and in the nature of the included organisms at different horizons and in different localities. In the case of so complex an assemblage of deposits, the explanation of their origin is also in all probability complex.

The fluviatile explanation has been adopted by Barrell[1] and his arguments are very strongly in favour of this explanation for parts at any rate of the deposits. They are based largely on the character of the conglomerates, and on sun-cracks and other features on bedding surfaces, indicating frequent exposure of the surface to the air. Barrell's views are of so high an interest, that a short summary of them may be given here. He specially advocates fluviatile deposition on account of the extent, thickness and coarseness of the conglomerates. He points out that thick and extensive pebble deposits are being laid down by torrents at the present day at the foot of mountain ranges, and on the flood plains in the upper parts of river basins. No thick marine gravels are known, the maximum being about 100 ft. The same is shown by the study of ancient marine conglomerates. Modern pebble-beds to the thickness of thousands of feet may accumulate at the foot of mountain ranges, as for example in the foot-hills of the Himalayas. Stream-laid gravels may be carried far and wide over a flood

[1] Barrell, J., *Bull. Geol. Soc. Amer.* 1916, p. 345; Gregory, H. E., *Amer. Journ. Sci.* 1915, p. 487.

plain and thus form very extensive deposits, but subaqueous pebble-beds of marine or lacustrine origin are of very limited extent, since the currents and waves are unable to carry coarse material far from the shore and tend to deposit it in layers parallel to the coast. The last argument however does not carry so much weight as the former ones, for the belt of pebble deposits shifts its position during the process of emergence or submergence, thus giving rise to widespread sheets of pebbles.

The formation of some of the deposits however, in tracts of water, is clear, though whether these were lacustrine, brackish water or marine is a moot point. Sir Archibald Geikie, as is well known, advocated the formation of the British Old Red Sandstone in five lakes, to which he gave the names of Orcadic, Lorne, Caledonia, Cheviot and Welsh Lake. These names may well be retained as geographical expressions, even if they prove not to have been lakes in the strict sense. Whatever may have been the actual mode of formation of the deposits, there seems to be no doubt that they were laid down in constricted or partly constricted sheets of water of limited extent, having a general S.W.—N.E. trend, and at some periods their shorelines approximated to the margins of the present outcrops, while at others as the result of earth-movements the waters were more widespread, causing overlap, at others again more restricted, causing unconformity. The conditions were favourable for the formation of deposits of an anomalous character as compared with those ordinarily formed in the open ocean. One of the arguments in favour of a lacustrine origin was based on the predominant red colouration of the beds, due to deposition of pellicles of ferric oxide around the constituent grains. The iron compounds were probably brought down by rivers draining an area of igneous and metamorphic rocks, but it is not clear how they were precipitated. At any rate the red colouration is not of great importance as giving definite indications of the lacustrine or marine nature of the water tracts. The formation of red deposits in the present ocean off the coast of Brazil has already been mentioned, but we need not go further afield than our own deposits of Devonian time in order to show that red deposits were formed in the sea, and, conversely, those of other colour in one of Geikie's lake tracts, continued into Carboniferous times. It is well known that the Old Red deposits of the Welsh Lake do include sediments with a normal marine fauna (the Skrinkle Sandstone). Further the red Hangman Grits of North Devon, and the similarly coloured Lincombe and Warberry Beds of South Devon are marked by the inclusion of normal marine faunas. On the other hand, among the deposits of Geikie's lakes are some not exhibiting red colouration, such as the Caithness Flagstones and certain yellow sandstones. These, it is true, are among those of doubtful origin, but in the Calciferous Sandstone of the Scottish central lowlands deposited in "Lake Caledonia," which continued into Lower Carboniferous times, there are deposits not of a red colour, containing a definite freshwater fauna. The red colouration is therefore not conclusive as indicating a freshwater or marine origin, though as above maintained it is more likely to occur in restricted waters, though not invariably so.

It is to the organisms that we must turn for the best evidence as to the origin of the deposits, but here again we are confronted with difficulties. The general absence of invertebrates secreting calcareous parts, and the abundance of Eurypterids and of the characteristic Old Red fishes have been adduced as an indication of the exceptional conditions of formation of the deposits, as no doubt they are to some extent. The general absence of lime-secreting invertebrates points to exceptional conditions, but by no means necessarily to the freshwater nature of the deposits, for such conditions are well known to have occurred in certain marine tracts, as will be discussed subsequently.

The actual occurrence of the fossils which are found in the Old Red Sandstone is of more importance, though even these give an uncertain note. The fishes may have lived in fresh water, but there is no definite proof of this, for they certainly did inhabit the sea, as shown by their presence associated with ordinary marine forms in open-water marine beds of Devonian age in Bohemia, and they may therefore have been marine only. As regards the Eurypterids Dr Marjorie O'Connell has very ably advocated their exclusively freshwater habitat[1]. She recognises the existence of Eurypterid remains in various open-water marine

[1] O'Connell, M., *Bull. Buffalo Soc. Nat. Hist.* vol. II, No. 3, 1916, p. 1.

sediments, as those of Bohemia, but points out that
these are always fragmentary and have been carried
into the sea from their freshwater habitat. Her con-
tentions certainly support very strongly the freshwater nature of their habitat, but I am not convinced
that a marine existence was altogether impossible for
them.

Of other organisms, some have been claimed as
freshwater, others as marine, in each case with apparent justification. The Crustacean *Estheria* with a
horny test may have lived in water of very various
degrees of salinity, being apparently very adaptable,
but some undoubtedly marine animals have been
found, especially a *Conularia* in the Old Red Sandstone of Herefordshire, and some *Leperditiae* in red
shales of the Lower Old Red of Carmarthenshire,
though these latter may also have been adaptable to
diverse conditions. The marine fossils of the Skrinkle
Sandstone have already been noticed. To set against
all these is the bivalve *Archanodon*, generally regarded
as closely related to the modern freshwater mussel,
Anodon.

If the determination of the habitats of the above-
named organisms is correct, it merely indicates that
certain of the beds of the Old Red as a whole are of
marine, others of freshwater, origin, leaving a large
proportion of uncertain origin.

The Rhynie chert should be mentioned here though
its characters are so exceptional and its distribution
so limited that it cannot be regarded as throwing
much light on the origin of the Old Red Sandstone

in general. It occurs in the Middle (or possibly Lower) Old Red of Aberdeenshire and resembles a siliceous sinter. It is obviously of marsh and aquatic formation and the organisms strongly suggest fresh water; some of them are terrestrial, but there is abundance of remains of a Phyllopod Crustacean, *Lepidocaris*, which is similar to the living freshwater *Branchipus*[1]: though the deposit consists of alternate layers of peat and sandstone, the silicification being of later date.

It would seem then that the Old Red Sandstone was not formed all in one way, but that its origin was complex. Some portions were probably fluviatile and lacustrine, others of marine origin, and the evidence on the whole is in favour of the formation of the latter in constricted water areas, with conditions not altogether dissimilar in some respects to those of the present-day Baltic, which has a restricted fauna and shows intermingling of freshwater and marine shells in the most brackish portions, and just as the Baltic, as a result of tectonic changes, has recently altered its character, the *Littorina* Sea and the *Ancylus* Lake having preceded its present condition, in the Old Red water areas somewhat analogous variations may have occurred.

[1] See Kidston and Lang, *Trans. Roy. Soc. Edin.* vol. LII, pt IV, 1921, p. 892; Horne, J., *Scottish Geogr. Mag.* vol. XXXIII, 1917, p. 385.

CHAPTER X

THE MUD BELT

THE mud belt, save under exceptional conditions, is found as a strip, generally of a width considerably exceeding that of the belt of variables. The inner mud line, separating it from the belt of variables, is apt to be irregular, the outer mud line more regular, owing to the greater regularity of the sediment-distributing currents far out at sea than of those in the neighbourhood of the coast. In the case of the larger areas of partially restricted sea, the mud belt may be continued into the restricted area as a tongue from the main belt, or even as an isolated outlier. This is also true in some cases of the organic belt.

The essential component of the deposits of the belt is of course mud, with a certain amount of sandy matter on the shoreward side, causing silts to predominate there, forming part of a transition strip into the belt of variables. Farther out to sea the mud will be finer and finer.

Apart from the mud particles little or no mechanically transported sediment will be found, save in exceptional circumstances, such as transport of coarser material by floating vegetation or ice. The majority of the objects larger than mud particles will be due either to concretionary and pseudo-concretionary action, producing bodies such as phosphatic nodules and glauconite grains, or to the remains of the hard parts of organisms. These may be sparsely scattered through the mud, or may occur in considerable abundance, causing addition of calcareous, phosphatic or siliceous material to the mud. As the number of organic remains increases, the deposits gradually pass from pure to calcareous mud, to argillaceous limestone and finally to pure limestone. Such changes occur mainly near the outer mud line, but also locally in the mud belt itself, owing to a temporary diminution in the supply of mud, or its complete cessation. In the latter case the mud belt of the area affected is temporarily replaced by the organic belt, but as the change is temporary and local, a consideration of such deposits is naturally included in this chapter.

A similar concentration of organic material, as already stated, may be produced by winnowing action, causing the removal of the whole or a part of the mud, leaving the organic residue as a more or less pure deposit. Such action will be less marked here than in the belt of variables, owing to the slighter power of the currents, but study of the rocks of the geological column shows that it does occur.

The varieties of mud will be considered in the same order as those of sand, though it is desirable to add another variant of the terrigenous muds to those suggested by Sir John Murray.

The term *mudstone* is sometimes applied to such muddy deposits as are not laminated, as distinguished from shales. Strictly speaking the term should be applied to all mud-formed strata, whether laminated or not. The subject of lamination has already been briefly alluded to, and it is probable that lamination is due to more than one cause. Lamination is in many cases no doubt due to pauses in deposition, but there are frequent cases where it is hard to say why it is or is not present. Many belts of argillaceous rock apparently formed under conditions generally similar differ much in respect of laminated structure: contrast for instance the usually unlaminated Oxford Clay of Britain with the frequently laminated Kimeridge Clay. A common type of laminated structure is due to abundance of flat objects, as for instance mica flakes and graptolites.

The striped structure known as "varved," though most marked in the silts of the belt of variables, may also be found among the deposits of the mud belt, where the stripes are very much thinner. An example is furnished by some of the Wenlock Shales of Great Britain, where the stripes are much more numerous to the inch than in the case of the Bannisdale Slates of the belt of variables, which have already been noticed.

Ordinary Terrigenous or Blue Muds. The general characters and composition of the muds have been fully described by Sir John Murray and are well known: they need not be enlarged upon here. It may merely be stated that the muds are found to consist of mineral chips and flakes of similar composition to those of the sands, but of smaller size and in different proportions. With these is mixed a variable amount of finely divided amorphous matter, probably largely silicate of alumina. In the ordinary muds much of the soluble matter of the original component grains has no doubt been leached out, giving the muds a higher alumina percentage than the sands, but in some circumstances a smaller amount of alumina and more soluble substance than that characteristic of the normal muds may be present, as for instance when the material deposited in the mud belt has been brought down by rivers traversing desert regions, or where an abnormal amount of volcanic dust is mixed with the ordinary terrigenous material resulting from erosion.

The colouring matter of the blue muds is stated by Murray to be due to organic matter, but in some cases it seems to be due to the presence of iron compounds.

Organisms in Blue Mud. Micro-organisms of planktonic habit, such as Foraminifera and Radiolaria, are often found in muds, sometimes in great quantity, so that the muds may pass imperceptibly into organic deposits. Sponge spicules are also frequently found in abundance. The larger organisms vary considerably as regards frequency, some muds of the belt being practically unfossiliferous, others containing fossils in great numbers. The muds probably contain more organic remains when accumulated in fairly shallow water than when laid down in the depths, for conditions such as light, temperature, food, and aeration of the waters are more likely to be favourable to organic existence in the shallows than in the depths, and the bathymetrical range of the muds is great. The relative abundance of organisms in deposits laid

down in clear and muddy waters respectively varies
considerably, but as regards number of species it may
be stated that the faunas of waters sullied with mud,
though individuals may be abundant, appear to be on
the whole poorer in species than those of clear waters,
and this will apply particularly though not exclusively
to the benthonic forms. Accordingly many of the
faunas of the ancient muds are not marked by great
variety of species; this is true in the case of some of
the ordinary blue muds, and, as we shall see, is even
more marked in the case of certain variants.

The organic contents of the muds may be partly
planktonic, partly benthonic. Even if the surface
waters charged with mud are unfavourable to the
existence of planktonic creatures, their dead tests may
be floated from elsewhere by surface currents, as is
believed to have been the case with many ammonites,
and accordingly we may find abundance of such in
deposits in which benthonic forms are few in species
and perhaps even in individuals.

To ascertain the conditions of deposition of some
of the muds (and this applies to others in addition
to the blue muds) it is important that we should be
able to distinguish the benthonic organisms from the
planktonic. This can be easily done when we are
dealing with existing forms and their near allies, but
in the more ancient rocks where the organisms are
only distantly related to living forms it is more difficult. Certain modifications of structure may however
serve as useful indices. Many planktonic crustacea for
example seem to have loosely knit carapaces, as com-

pared with their benthonic relatives. Dr L. Dollo has suggested that some of the Palæozoic trilobites had carapaces modified to enable them to live under planktonic conditions. Nevertheless in the same group of organisms we get almost certain indications of benthonic habitat. Some of them have exceptionally flat tests which would enable them to lie concealed on a muddy floor, and that this was the purpose of this modification is indicated by the occurrence of the forms with flat tests in great abundance in the muddy deposits. Attention has frequently been called to the fact that many trilobites are blind whereas others have very large eyes with abundant lenses: this is usually considered to indicate a deep water habitat, but is probably also an indication of benthonic conditions, for even if the eyes be regarded as having been modified on account of absence of sunlight in the depths, creatures living at such depths were probably benthonic. But the modification of the eyes may have been on account of the gloom produced by the mud-sullied waters. Another modification of the eyes of trilobites is even more suggestive of this, especially as it occurs in forms with flat tests, such as *Arethusina*. Here the eyes, instead of being raised and reniform to allow of all-round vision, horizontal as well as upward, are modified to permit upward vision only; being of a circular form with few lenses, all directed upwards and therefore almost certainly borne by a creature living at the bottom, and only requiring overhead vision.

There is still much room for study of the structures

of extinct groups of organisms as bearing upon the conditions in which they existed.

Green Muds. The muds whose colour is due to the occurrence of glauconite grains are similar in character to the green sands, which owe their colour to the same cause, the glauconite grains having no doubt been formed under similar conditions in the two cases, except with regard to the size of the grains of terrigenous material in which the glauconite grains are embedded. Their origin therefore need not be further discussed, though it must be admitted that much obscurity still surrounds it.

Glauconite mud seems to be more abundant in the belt of variables, but it does occur in the mud belt, as for example the glauconite muds which are well developed in the Gault of England, clearly, from its distribution deposited in the mud belt and not in that of variables.

It was previously pointed out that many green muds owe their colour to a substance other than glauconite, though in certain cases at any rate it may be to some extent comparable with it, as regards mode of formation. This colouring matter appears to be essentially silicate of iron, though it is desirable that further investigations should be made concerning it. These green muds coloured by iron silicate are abundant in the Lower Palæozoic rocks of Britain, for instance in the Llanberis Slates of the Lower Cambrian and the Tarannon Shales of the Valentian. There is reason to suppose that they are essentially similar, as regards origin, to certain muds of a different colour,

which would have been green but for the presence
of a more intense colour due to organic agencies. This
and other considerations favour the view that the
type of green mud now being considered owes its
character as regards colour to the general absence of
organic matter, due to its having been formed under
conditions unfavourable to abundant life, for fossils
are often absent or very rare in these deposits, and
when found they are often stunted, whereas associated
sediments of precisely similar character, save for
colour, are marked by abundance of organic material.

Green muds are certainly formed also in other ways,
as for instance the green bands often found intercalated with the red muds of the Keuper Marl, but
with these we are not concerned here: they are mostly
of non-marine origin, and if any of them be marine,
they belong to the belt of variables, and probably to
such parts as formed constricted areas.

Red Muds. These are formed under various conditions, some as we have seen, such as the Keuper
Marls, being indicative of continental conditions.
Those with which we are here concerned are laid down
in the sea. Such are those described by Murray as
being formed off the mouths of some of the great
South American rivers flowing into the Atlantic.
These drain an area with a hot moist climate and
extensive forests. Moreover igneous and metamorphic
rocks, often of a highly ferruginous character, are extensively developed. Therefore, as suggested by Raymond, the weathering will be of the lateritic type,
and the iron carried down by the rivers will be in a

highly oxidised condition. He also states that in his opinion the red iron compound finally deposited with the mud is not necessarily the anhydrous ferric oxide, but a red hydrated oxide[1].

Red deposits of a type similar to those now forming in the Atlantic as just described are found among the older rocks, and it is interesting to note that they are associated with the green deposits already described, namely those of Llanberis Slate and Tarannon ages. The Caerfai beds of St David's contain a band of red mudstone 50 ft. thick intercalated between two thick masses of sandstone, no doubt indicating a temporary encroachment of the mud belt over the belt of variables. A similar red mud is found in beds of the same age in North Wales, where it is more highly cleaved, so that any organisms preserved in it would escape detection. But the Caerfai mudstones do contain fossils, though these are very rare and extremely small, indicating the prevalence of conditions unfavourable to life.

An almost precisely similar deposit is found at the top of the Browgill (= Tarannon) Shales of Westmorland, again associated with green mudstones, which it immediately overlies. The scanty and dwarf fauna here also suggests conditions unfavourable to life.

In the case of the Welsh deposits just mentioned

[1] Raymond, P. E., *Amer. Journ. Sci.* vol. XIII, 1927, p. 234. See also Hager, D. S., *Bull. Amer. Assoc. Petroleum Geol.* vol. XII, 1928, p. 901. The paper last quoted contains a very full general discussion of the nature of the colouring matter of sediments.

it is certain that extensive areas of igneous and metamorphic rocks of Pre-Cambrian date existed on the
land surface adjoining the Cambrian sea, and that
this was the case over very wide areas in early Cambrian times, as we should expect, is indicated by the
widespread occurrence of red mudstones of *Olenellus*
age, as in Shropshire and Newfoundland. In the case
of the red Browgill beds of Westmorland the existence of a neighbouring land formed of such rocks
is not so certain, though probable.

It seems probable that the red muds hitherto described and the green muds often associated with them,
which latter are distinct from the glauconitic green
muds, were deposited in water tracts where the conditions were unfavourable to life. There are however
other red muds, seemingly of different origin, which
must now be considered.

Dr T. G. Bonney[1] suggested that the red colouring
matter of the Hunstanton Red Rock of Norfolk was
due to the decomposition of glauconite. The Gault
Clay gradually thins away northward and ultimately
passes laterally into the lower part of the Red Rock.
The Gault itself, as already stated, contains abundant
glauconite, and a certain amount is present in the
Red Rock, as shown by microscopic examination.
This Red Rock is an open sea deposit, and is a calcareous mud with a rich fauna. The glauconitic origin
of the colouring matter has certainly not been established, but the suggestion is plausible. The matter
requires further study.

[1] Bonney, T. G., *Cambridgeshire Geology*, 1875, p. 76.

Very similar deposits are found among much more ancient strata. In Sweden the Lower Ordovician contains a series of rocks known as the Orthoceras Limestones: some of these are grey, others red. The grey variety is in some places abundantly glauconitic, and the association of red rocks, extraordinarily like those of Hunstanton in lithological character, with others bearing glauconite suggests that the two sets of deposits were formed under similar conditions. The Orthoceras Limestone has an abundant and varied fauna, and there are no indications that the conditions in the sea in those times were inimical to life. Another red deposit also occurs in the Ordovician rocks of Sweden at a higher level, namely, the upper part of the Trinucleus Shales, which are red in some areas, as in Westrogothia. These are true muds rather than argillaceous limestones, and lithologically resemble the red beds of Westmorland much more closely than the Hunstanton Red Rock. The resemblance to the former is indeed very great; nevertheless the fauna is a rich one, and though the conditions of formation are obscure, it may be suspected that they were more nearly allied to those of the Hunstanton Rock and Orthoceras Limestone than to those of the Caerfai and Browgill beds.

Black Muds. There is every gradation from the ordinary blue terrigenous muds into black muds, indeed many sediments which are black do not show any other peculiarities. Some of these, occurring in the belt of variables, are evidently due to erosion and re-deposit of pre-existing black muds: an example is afforded by the black mud now forming in the estuary of the Seiont, already noted. Among ancient strata the black mudstones of the Lias seem to furnish an example on a larger scale, having been largely derived from dark Carboniferous shales. Nothing further need be said of these. The black muds requiring particular notice are specially developed in the mud belt: a striking feature of these is the rarity of benthonic organisms secreting calcareous hard parts, though those with siliceous hard parts and planktonic (or pseudoplanktonic) organisms may be present in abundance. It is obvious that conditions of exceptional character are indicated in these cases.

The Black Sea. This large, nearly land-locked, sea presents features of special interest to the geologist, on account of its wide area, great depth, and the peculiarity of the conditions that prevail in it. The central portions are at present receiving a deposit of black mud, for we are here dealing with an outlier of the true mud belt, in this instance that of the Mediterranean. I have spoken of its importance as throwing light on certain ancient deposits of black mud in the paper quoted below[1].

"It is interesting to note that this inland sea has a maximum depth of 1227 fathoms, which may truly be termed abysmal. The floor is covered with black mud, which receives much plankton from the surface waters. Bacteria flourish on the decomposing organic matter, and their presence accounts for the formation

[1] Marr, J. E., *Quart. Journ. Geol. Soc.* vol. LXXXI, 1925, p. 125.

of sulphuretted hydrogen, which poisons the lower waters and forbids the existence of benthonic life. Life is, therefore, limited to the disturbed upper waters above the 100 fathom line: below this lie the stagnant poisoned waters."

Posidonomya Beds of Wales. The lagoonal formation of certain limestones in the Lower Carboniferous of South Wales has already been considered. In the paper in which these are described Mr Dixon gives reasons for regarding the muds and associated deposits of the *Posidonomya* beds which occur between the Lower and Upper Carboniferous rocks as also of lagoon origin[1]. Some of these beds are black muds, others lighter in colour, but all characterised by a paucity of organisms secreting thick calcareous tests, and Mr Dixon rightly argues that this indicates the occurrence of exceptional conditions during the formation of the deposits. It is almost certain that these conditions prevailed over a much wider area than South Wales, for similar deposits belonging to this period are found in Devonshire, the Isle of Man, Yorkshire, and in many parts of the Continent. Nevertheless the evidence is in favour of deposition in a constricted water area though of wide extent. That the conditions were those of shallow water is clearly shown by the evidence put forward by Mr Dixon. What caused terrigenous sediment to be deposited in what had been a clear sea in Lower Carboniferous times is not apparent, but whatever the causes, they were intermittent during the time of formation

[1] Dixon and Vaughan, *loc. cit.* on p. 163.

of the *Posidonomya* beds, for pure organic deposits are interstratified with those of terrigenous mud. In the mud beds of this group forms generally regarded as planktonic, secreting calcareous tests, are often abundant, especially Goniatites, but the benthonic forms with calcareous tests are not abundant, though quantities of siliceous organisms occur, planktonic radiolaria as well as benthonic sponges. These sometimes form variable proportions of the muds, at other times they occur nearly pure, as sponge cherts and radiolarian cherts. These cherts should be spoken of as belonging to the organic belt, but as there is every passage from the muds containing the siliceous organisms to the pure cherts, they may be considered here.

Radiolarian deposits have been regarded in the past as being always of abyssal origin, though as regards certain of them this view is no longer held, thanks largely to the work of Mr Dixon himself. They occur in the abysses of modern seas, on account of the prevalence there of the two conditions necessary for their formation, namely, absence of mechanical sediment and of calcareous material. Mechanical sediments are absent also from many of the shallower parts of the ocean, past and present, but the sea-floor is there usually occupied by calcareous material. This is absent from the modern abysses owing to the solution of calcareous tests before they reach the sea-floor. If there should be any other cause preventing accumulation of calcareous material there is nothing to hinder the formation of pure radiolarian deposits in the shallows. It does not appear that any such have yet been recorded in the modern seas, but when a more detailed examination of the floors of some of our constricted sea areas has been made, it is quite likely that they may be discovered. The question is, what was the bar to the formation of calcareous material in the case of the cherts of the *Posidonomya* beds? I believe that it was some iron compound, a view supported by the conditions prevalent in the Black Sea, and by the nature of the faunas of the graptolite-bearing black muds to be considered presently. It may be noted that nodules of ironstone are present in the *Posidonomya* beds of South Wales, and occur in similar beds elsewhere. Iron compounds, as before stated, are more likely to occur in appreciable quantities in restricted waters than in those of the open ocean, and accordingly we may expect that apart from the oceanic abysses, radiolarian deposits will be most readily formed in constricted waters away from the ocean centres. No doubt the clear surface waters far away from land are more favourable to the existence of radiolaria than the mud-sullied shallows, but a sufficient number can be introduced into shallows to furnish deposit. These may be brought in dead as well as in the living state, and are found even in estuarine waters. Furthermore, when the pure cherts of the *Posidonomya* beds were being formed their seas were as suitable for the existence of radiolaria as the actual waters of the open sea.

Graptolitic Mudstones. In a paper on the Stockdale Shales, to which reference has already been made, I discussed the modes of formation of those deposits, and

my conclusions as to their origin are quite in accordance
with those of Mr Dixon. The Stockdale Shales are of
Valentian age, and present what is usually spoken of
as the graptolitic facies, as opposed to the shelly
facies of contemporaneous beds elsewhere. They are
thin compared with the shelly beds, with a minimum
of about 200 ft., and they present four types, black,
blue, green and red. On the whole they occur in that
order, in vertical succession, with black at the base,
but one type is often found interstratified with another
in thin bands. Lithologically they are similar as regards their component terrigenous matter, differing
only in the colouring matter and the proportion and
nature of the organic material. The colouring matter
of the black muds is partly carbon, and partly iron
sulphide in a finely divided condition. About 6 per
cent. of carbon is found in some of the beds. The blue
beds are coloured by iron carbonate, probably due to
metasomatic replacement of calcium carbonate. The
green colour is due to iron silicate, and the red to iron
peroxide or hydroxide. The graptolites occur only in
the black muds, those of other colours containing
other organisms only and these but sparingly. The
benthonic forms are small, and evidently lived under
unfavourable conditions. Here then, as in other cases,
the presence of iron compounds in the water was unfavourable to the existence of lime-secreting organisms.

The three varieties just dealt with belong to types
described in the preceding sections, but are specially
noticed here on account of their close association with

the black muds, which we shall now proceed to describe. In the case of the Stockdale Shales, benthonic organisms are absent, all being planktonic or pseudo-planktonic: the latter are the graptolites themselves, as shown by the researches of Prof. Lapworth. These often occur in countless profusion. Lapworth regarded them as living attached to floating weed, which furnished the carbon found in the mud. Iron sulphide is usually present in considerable quantity, both in the finely divided state and in concretionary masses. Though iron compounds in the water may have had their effect as deterrents of organic existence, the main factor was probably the sulphuretted hydrogen, developed much as it now is in the Black Sea, the decomposing algal matter being here the cause of development of bacteria. In the case of the Stockdale Shales all benthonic life (apart from the postulated bacteria) appears to have been absent, the water having been poisoned to such an extent that even organisms without calcareous parts could not exist.

It would appear then that the partly land-locked gulfs in which these muds were formed, according to the evidence derived from the study of their geographical distribution, were occupied by poisoned waters in the parts which were not aerated by surface circulation; that is, the conditions were analogous to those of the Black Sea, and planktonic organisms lived in the aerated waters near the surface, while their tests after death sank to the floor. This appears to have happened in the case of the graptolitic mudstones, for though benthonic forms are there absent

thin-tested floating animals like Phyllopods and
Cephalopods are found associated with the graptolites.
If the above argument be correct benthonic forms
should occur in the deposits adjoining the shores, and
above the 100 fathom line, which is taken as the upper
limit of the poison belt, and this is precisely what
happens. The physical evidence points to the Stockdale Shales and other graptolitic deposits of Britain
having been formed in a circumscribed gulf, whose
shores lay near the southern part of the Scottish
Highlands, and on or near the Welsh borders respectively. In the central deeps between these the black
graptolitic muds occur. As the shore is approached,
for example near Rhayader, the two types become
intermingled. In the belt of variables found in the
shallows and therefore in thoroughly aerated waters
a perfectly normal benthonic fauna is found, with
occasional fragments of graptolites drifted from the
tracts further away from the coast.

The conditions under which the black mudstones
of the Stockdale Shales were formed were probably
similar, in a general way, to those under which the
corresponding muds of other times and places were
formed, with slight differences due to local variations;
such differences seem to be indicated in the case of
certain deposits described by Mr King, which suggest
formation in a subsidiary constricted area within the
main gulf. This minor area lay between land in the
neighbourhood of Church Stretton to the east, and
sandbanks near Bala, which separated it from the
main gulf. In this partially confined tract the grapto-

lites, as well as the representatives of the shelly fauna, are dwarfed. These deposits were probably continuous with those of the main gulf, but of the transitional type found elsewhere between the main deeps and shallows[1].

In the Stockdale Shales no radiolaria have been found: in other areas however we meet them in graptolitic deposits of this and other ages, as for example in Saxony, where they are also Valentian. The most interesting example is furnished by the Glenkiln (Ordovician) Shales of the Southern Uplands of Scotland. These are sooty black muds, containing in the blackest part nothing but graptolites, unless perchance radiolarian tests are embedded in the black muds also and concealed by the colouring matter. The lowest division however of the Glenkiln series consists of radiolarian cherts; so that here again these cherts are found associated with black muds containing forms living under exceptional conditions.

The many similarities between the *Posidonomya* deposits of Carboniferous age and the Lower Palæozoic graptolitic mudstones are of interest, considering that each has been regarded independently as having been formed under conditions necessary for the production of "poisoned" waters, which are here believed to have been due to stagnation in the deeper parts of partly land-locked areas, already compared with those of the Black Sea. It is conceivable however that similar conditions may exist in parts of the open ocean itself,

[1] King, W. B. R., *Quart. Journ. Geol. Soc.* vol. LXXIX, 1923, p. 487.

and may have also done so in the past, where circumstances are favourable to production of bacteria in the deeps. Such might well be brought about by sinking of algal matter to the floor where algae formed dense floating masses at the surface, as in the case of the Sargasso Sea. It may be wished that direct information on this point could be obtained, but its practicability may be doubted, owing to the difficulty of sounding and obtaining floor-samples in such areas.

Volcanic and Coral Muds. As the volcanic sands and coral sands have already been considered, it is unnecessary to treat of the corresponding muds at length, since they differ only in size of particles; they are found frequently in the rocks of past ages, whenever vulcanicity was rife nearby, or when conditions were favourable for the existence of reef-corals.

CHAPTER XI

THE ORGANIC BELT

BEYOND the outer mud line lies the *Organic Belt*, extending outwards to the area of abyssal clay deposits. The outer mud line as previously noticed is not a line, but a transitional belt of varying width consisting of mud and organic matter, the former most abundant on the inner, the latter on the outer side.

Before describing the features of the organic belt, it may be well at this point to say something concerning variations in vertical sequence. Certain deposits of organic origin have so wide a lateral distribution that they might be regarded as belonging to the organic belt, yet when we study the great thickness of the strata to which they belong we may find that they really form components of a gigantic belt of variables, if we regard time as well as space, and this may be preceded and succeeded by deposits of the mud belt. This may be illustrated by reference to the British Jurassic rocks alluded to in a previous chapter. It was there stated that the Lias, Oxford Clay and Kimeridge Clay indicate formation in the mud belt; between the two first named the Bajocian and Bathonian rocks belong to the belt of variables if we take time as well as space into account, as do also the Corallian rocks between the two Oxford and Kimeridge Clays, and the Portland and Purbeck rocks above the latter.

The clays have a fairly uniform character along the whole of their outcrop in England, whereas the other deposits do vary laterally as well as vertically. They are therefore relegated to the belt of variables, but if we regard only the short period of the formation of the deposits of one of these belts, for instance, that of the rocks between the Lias and Oxford Clay, we may get a sequence from the variable to the organic belt existing during a comparatively limited period of time; for example, part of the variable deposits of the Bathonian as found in the north of England pass into deposits of the mud belt, namely, the Great Oolite Clay, as we travel southwards, and proceeding still farther south the equivalent calcareous deposits of the organic belt in south-west England form a portion of the Bath Oolites. It is in fact a question of degree: just as in space sub-belts may be traced locally in the belt of variables, so also they may be traced in time, and no sharp line can be drawn between the sub-belts of various sizes in the belt of variables and the major belts themselves. In the case of the British Cretaceous rocks we do seem to be dealing with actual belts rather than sub-belts. The Lower Cretaceous marine deposits belong to the belt of variables in the south but in Yorkshire to the mud belt (Speeton Clay). If the organic belt existed its deposits now lie under the North Sea. Later, as the sea encroached southwards during subsidence, the belts correspondingly shifted; and in Gault times the belt of variables is found in the extreme south of England: as one passes northward it changes into the mud belt, marked by the Gault Clay, resting on the deposits of the belt of variables of Lower Cretaceous age, and in Lincolnshire and Yorkshire this clay passes laterally into part of the Red Chalk, belonging to the organic belt, which in the north rests upon the mud belt of Lower Cretaceous times. Still further change is noted in the case of the Chalk; subsidence had then gone on to such an extent that the English deposits belong exclusively to the organic belt.

It is obvious that conditions will be more favourable to a clear demarcation of the different belts in the case of truly open-sea deposits than in those of partly land-locked areas, many of which as already seen are occupied only by the belt of variables, though in the larger of these the mud belt may occupy the central parts, as already seen. In favourable circumstances the organic belt may also be found in these constricted areas, but apparently only in those of considerable size, for mud, as has been seen, can be carried 100 or even 200 miles from the coast, and a constricted sea with a shorter diameter of 200 miles may therefore be completely filled by terrigenous deposits, especially if sedimentation keeps pace with subsidence.

It has been pointed out that N.W. Europe may possibly never have been occupied by open sea of great extent in past times, but that the observed phenomena are explicable on the supposition that the marine areas of that region were more or less constricted. In such circumstances it is doubtful whether any important deposits belonging to the organic belt were formed save under exceptional conditions. Two such deposits do however occur, namely, in the Mountain Limestone and the Chalk. These have been partly considered, but more remains to be said. The marked similarities of character are not accidental, but must betoken generally similar conditions of formation, which, it has been suggested, indicate the shallows of warm-water tracts perhaps lying off lands under arid conditions[1]. The probability is that the wide distribution and thickness of nearly pure calcareous material are due to the practical absence of the mud belt, and the extreme narrowness of the belt of variables along many parts of the coast, for in the case of both the Mountain Limestone and the Chalk the actual littoral deposits hugging the coastline are sometimes found to be formed of pebbles in a calcareous matrix, indicating clear water up to the coast itself. In the absence of the mud belt all the deposits, save the littoral ones just mentioned, must be regarded as belonging to the organic belt.

There seems to be full reason for concluding that both the Mountain Limestone and the Chalk were laid down in shallow water, and not as abyssal deposits, though the latter origin has been assigned to the Chalk. Mr Dixon has given ample reasons for attributing a shallow-water origin to the Mountain Limestone, and A. R. Wallace, in his *Island Life*, has argued at great length and convincingly concerning the shallow-water origin of the Chalk: he has given a general indication of the margins of the Chalk Sea. It is interesting to note that he believed the con-

[1] See Bailey, E. B., *loc. cit.*

ditions of deposition to approach most nearly to those now existing in the subtropical seas of the North American Atlantic region, mentioning specially the seas off the Bahamas, which, as we have previously seen, show very special relationship with the Chalk as regards the lithological characters of the sediments now being formed there. As in the case of the recent subtropical Atlantic deposits, so in the Mountain Limestone and Chalk, marginal coral reefs occur; in the last named they appear to be uncommon, though recorded in that of Faxoe, Denmark. Wallace discusses the bearing of the Chalk fauna on the depth of water, and points to its shallow-water nature, as borne out by its resemblance to the fauna of the existing shallows and also by the absence of typical deep-sea forms. He believes that the abundant foraminifera (including *Globigerina*) may have been carried into the Chalk gulf by a Cretaceous "Gulf Stream," prevented from flowing into arctic tracts by a land-barrier extending south-eastwards from Greenland. It may be taken then that the Chalk is in no way an abyssal deposit; this is what we should expect, considering the rarity of even deposits of the shallower parts of the organic belt in N.W. Europe. In such conditions as seem to have prevailed there during geological ages it is unlikely that true abyssal deposits should be found. They must be searched for elsewhere.

In the organic belt siliceous as well as calcareous organisms may be found. This is the case with many modern deposits, and also in those of ancient date, of which the two deposits just described furnish examples. There is little doubt that many of the chert bands of the Mountain Limestone and the flint bands of the Chalk have a similar origin, and in both cases sponges seem to have played a prominent part in their formation. There is an interesting fact in connection with the flints of the Chalk; occasionally a flint nodule is found, which is not solid throughout, but contains some of the original Chalk matrix securely sealed in its interior: this "flint-meal" has therefore escaped alteration, and a sample has been subjected to careful scrutiny by Dr G. J. Hinde[1]. The matrix thus entrapped consists of ordinary Chalk material mixed with a large quantity of sponge spicules, and it is therefore clear that those parts of the Chalk in which the flint lay had a similar composition, but that the sponge spicules have been removed as the result of subsequent chemical changes. This gives a useful warning, when comparing ancient with modern deposits, that the former need not necessarily now be in the state in which they were originally deposited.

From what has been said it appears that the deposits of the organic belt cannot be very extensively studied in N.W. Europe. Their full investigation must be carried out elsewhere.

The Organic Belt in Existing Seas. Comparatively little seems to be known in detail about the non-abyssal deposits of the open waters of the modern oceans, though researches into this matter are now being carried on. It may be noted that Murray's

[1] Hinde, G. J., *Fossil Sponge Spicules from the Upper Chalk,* Inaug. Diss., Munich, 1880.

Pteropod Ooze might well be regarded as belonging to this category, as it is not found in truly abyssal depths. No doubt in many places the mud belt is immediately succeeded by the abyssal oozes without intervention of any organic deposits formed in shallow water. We have already noticed the latter type of deposit as occurring in exceptional cases, such as the shallows of the subtropical waters of the American Atlantic coast, but much remains to be done in the study of other shallow-water areas where organic deposits lie outside the mud belt. In the meantime the deposits of this character formed in past ages afford much information, which will give material aid in the study of recent deposits.

One might divide the deposits of the organic belt into two sub-belts, non-abyssal and abyssal, but the line of demarcation would be hard to draw, and as was before noted, and will be enlarged on later, owing to the rarity or possible absence of abyssal deposits among the rocks of the past, such a division would be of little value to the geologist.

No doubt the recent shallow-water deposits outside the mud belt generally contain a mixture of benthonic and planktonic forms, but there will be great variation in detail, and with our present limited knowledge of these deposits it will be of little use to discuss them at length.

Ancient Deposits of the Organic Belt. It has been argued that these deposits occur rarely in N.W. Europe and were only formed under exceptional conditions, such as controlled the deposit of the

Mountain Limestone and the Chalk. To obtain a fuller knowledge of their characters other areas must be studied, especially those which show evidence of wide areas of deposition in the great geosynclinal regions of past times. As an example of these we may take the region called Tethys by E. Suess: this was bounded by Gondwanaland on the south and Angaraland on the north. It seems to have been a region of deposition under somewhat similar physical conditions with but slight interruption, at any rate from Permo-Carboniferous to Eocene times inclusive. In its central parts, outside the mud belt, vast deposits of limestone were formed, of great extent and thickness, and these may be regarded as normal, as contrasted with the abnormal calcareous gulf deposits of N.W. Europe. An almost continuous deposition of calcareous sediment in Tethys may be traced from the Cephalopod-bearing limestones of Permian, Triassic and Jurassic ages, through the Hippurite limestones of the Cretaceous to the Nummulite limestones of early Tertiary age. We need not enter into details as to minor characters of the limestones: the main difference between those of different ages seems to lie in the character of their faunas. These are often very rich, and incidently afford valuable aids to correlation. The limestones are often very pure, and frequently white, owing to the small proportion of terrigenous matter. Their faunal assemblages indicate that they are not of abyssal origin; their general uniformity of character, maintained in deposits which represent long periods of time, taken in conjunction with their similar uniformity over wide areas, would show that we are here dealing with shallow-water deposits of a truly open ocean of wide extent and not with constricted areas such as those supposed to have existed in N.W. Europe.

One result of the conditions differing in the open ocean and constricted seas is most important, and may now be considered. The conditions prevalent where the organic deposits were laid down in the open ocean, in which subsidence was accompanied by deposition keeping pace with it, led to the formation of very thick masses of sediment. The formation of this was not accompanied by winnowing of the finer material and the production of minor erosion planes, no one of which need be marked by an important palæontological break. The result is that one may meet with very thick deposits of pure limestone formed some distance from the land, representing quite thin sediments formed during the same period in the belt of variables, where the above-mentioned processes are in constant operation. As an instance of this the 580 ft. of the Bracklesham Beds of the Hampshire Basin may be contrasted with the much greater thickness of the equivalent part of the Nummulitic limestone, and the thickness of the whole Eocene succession of Britain is measurable by hundreds, as contrasted with the thousands of feet of the Eocene rocks of Tethys, although there is no marked palæontological break in the British sequence.

The deposits of the Tethys region have received much attention: those of other large geosynclinal areas await further study. Such is the Arcto-Pacific Sea of Moisisovics, comprising the Arctic regions and the Pacific belt to the south of them. Furthermore, a much fuller study than has yet been carried out remains for completion in the case of the deposits of this belt among the Palæozoic rocks. The work of the late Dr C. D. Walcott on the great limestone deposits of Lower Palæozoic age in the Canadian Rockies shows how much has yet to be done in other and little-known areas, for there is little doubt that many of these geosynclinals occupied areas of whose geology our knowledge is exceedingly scanty. These statements may act as a warning that a full knowledge of the sediments of N.W. Europe does not furnish the key to all the problems of sedimentation in past times.

Possible Abyssal Deposits of Ancient Date. A considerable difference of opinion existed and still exists as to the occurrence of true abyssal deposits among the sediments which have emerged to occupy land tracts. Dr A. R. Wallace, a strong supporter of the theory of the permanence of ocean basins, discussed the origin of the Chalk in support of it, and, as before stated, brought forward most important evidence as to the shallow-water origin of that deposit. Geologists, however, soon after the appearance of *Island Life*, claimed a large number of other ancient deposits as abyssal. Some of these were considered to be representative of the modern Globigerine Ooze, and many siliceous deposits formed of radiolaria were claimed as being ancient analogues of the modern Radio-

larian Ooze. Indeed, Dr G. J. Hinde seems to have
regarded all these radiolarian deposits as abyssal.
It has already been shown, however, that in the case
of some such as those of the *Posidonomya* beds of
Wales and elsewhere the evidence for shallow-water
origin is strong, indeed convincing, so that the
presence of radiolarian cherts cannot in itself be
considered proof of abyssal origin. It is therefore
desirable that the various instances of deposits claimed
as abyssal since the appearance of *Island Life* should
be more fully considered. The calcareous deposits may
first be mentioned and afterwards the siliceous. It
may be assumed that those who in recent years have
claimed the occurrence of deposits analogous to the
modern Globigerine Ooze were conversant with
Wallace's objections in the case of the Chalk, and
made a more critical examination of the problem
than did those who wrote before Wallace's criticism.
Two cases may be noted where sediments have been
definitely claimed as Globigerine Oozes. One is Mr
R. J. L. Guppy on Eocene beds of Trinidad[1], the other
Mr R. B. Newton on Cretaceous deposits in Madagascar[2]. In Mr Guppy's paper deposits varying in
age from Cretaceous to Pliocene are described, and
groups of sediment of alternate deep- and shallow-
water origin are claimed to occur; in beds newer than
the *Globigerina* limestones are radiolarian deposits.
There is no doubt that some of the Trinidad deposits

---

[1] Guppy, R. J. L., *Quart. Journ. Geol. Soc.* vol. XLVIII, 1892,
p. 519.

[2] Newton, R. B., *id.* vol. LI, 1895, p. 72.

are of deeper water origin than others, though it may be doubted whether any of them are truly abyssal. Mr Guppy's researches are, however, well worthy of study as illustrating certain exceptional modern deposits of marine origin. The abyssal nature of the Madagascar deposits, although possible, does not appear to be fully established.

The other modern calcareous deposit of this category is the Pteropod Ooze, whose truly abyssal origin we have already questioned. It is only a variant of the Globigerine Ooze. Pteropods often occur as fossils in considerable abundance and one deposit is known which is actually a pteropod limestone. It forms one of the members of the Hamilton (Devonian) series of Canada: it was mentioned and a microscope section of it figured by Prof. H. A. Nicholson in Nicholson and Lydekker's *Palæontology*. It is not, however, analogous to the Pteropod Ooze, not being a variant of the Globigerine Ooze, but composed essentially of pteropods only of the genus *Styliola*. It is a *Styliola* limestone. Judging from the accompanying sediments, it is a shallow-water deposit formed in the belt of variables.

As indicative of true abyssal origin of some of the ancient sediments most stress has been laid upon the siliceous deposits, especially those formed of radiolaria. Diatomaceous cherts and earths are found among various ancient rocks, but their abyssal origin has never, I believe, been seriously asserted. Their mere presence is no indication of abyssal or even deepwater origin, and at the present day diatomaceous deposits are actually being formed in lakes. Some of those of ancient date may also be lacustrine, others are certainly marine, but the associated strata point to their shallow-water origin.

There remain the radiolarian earths, cherts and jaspers. Some of these have already been noticed, namely, those occurring in the *Posidonomya* beds, and in certain graptolitic deposits, and their non-abyssal origin established. These are examples of numerous cases, widespread, and belonging to many geological periods, of which one other striking example may here be noted. This is found in the Devonian rocks of New South Wales, as described by Pittman and David[1]. The radiolarian deposits there are partly cherts, partly clays containing radiolaria in various proportions, associated with other deposits of obvious shallow-water character. The great thickness of these deposits (nearly 10,000 ft.), the occurrence of remains of land-plants in them, and other reasons induced the authors to assign a shallow-water origin to the whole, and it is difficult to see how this conclusion could be combated.

There is, however, one instance which stands by itself as regards the nature of the deposits associated with it, namely, the well-known radiolarian earth of Barbados. In this case evidence of shallow water is lacking, while on the other hand positive evidence points to deep-sea, if not actual abyssal conditions. Dr J. W. Gregory has described from the Scotland

[1] Pittman, E. F. and David, T. W. E., *Quart. Journ. Geol. Soc.* vol. LV, 1899, p. 16.

formation, a calcareous deposit in contact with the radiolarian earth, two echinids, *Cystechinus crassus* and *Archæopneustes abruptus*, which genera are now found living at depths of 2200 and 208 fathoms respectively. One has a somewhat thinner test than the form found in the Scotland formation, which causes Dr Gregory to suggest that the water in which the latter lived may have been somewhat shallower than that which forms the habitat of its living congener. The *Archæopneustes*, therefore, being obviously shallow water, indicates that that part of the deposit containing it is not of deep-sea origin, but it is found in a different part of the limestone to that which has yielded *Cystechinus*. The latter, though having a thicker test than the modern species, is regarded by Dr Gregory as definitely indicating deep-water origin of the containing sediment[1]. If the *Cystechinus*-bearing limestone be abyssal, there is no reason why the associated radiolarian deposits should not also be the same. This view is indeed taken by Dr Gregory. Taking the evidence as a whole it would appear that this Barbados earth approaches more closely in character to the modern abyssal ooze than any other of the deposits of past times which have hitherto been described.

It might be urged that if these deposits of late Tertiary (probably Pliocene) date have been brought up from the depths to form part of the land surface, there is more chance of abyssal deposits of earlier

[1] Gregory, J. W., *Quart. Journ. Geol. Soc.* vol. XLV, 1889, p. 163, and vol. XLVIII, 1892, p. 640.

date having emerged, and that the theory of the permanence of ocean basins, therefore, receives no support from the asserted absence of abyssal deposits now occurring on the land. It must be remembered, however, that the West Indies are in the septum of a great earth-wave of modern date, and the area is one peculiarly favourable to emergence of deep-sea deposits. On the whole one is inclined to regard this Barbados earth as, if not abyssal, at any rate of a character intermediate between the truly abyssal and the shallow open-sea deposits of the organic belt.

It would appear also from the meagre results obtained after prolonged examination of sediments of various ages and areas, that if abyssal organic deposits do occur among the strata now above sea-level, they are only rarely met with, and form a very small part of the total bulk of the sediments open to our inspection. The greater part of them must still be lying below the ocean depths.

A word may be said about the abyssal clay, usually spoken of as the Red Clay, though it is by no means invariably red. If the abyssal deposits of organic origin are rare among the strata now occurring on the land, this must be more so, if indeed it exists there at all. Even if it does occur, it is doubtful whether it would be detected, for most of the characters by which it could be identified are negative, and of the positive ones, the two of greatest significance would be incapable of application. One is the occurrence of vertebrate remains of extinct species mixed with existing forms in the modern clay: this is obviously useless in older rocks where all are extinct. The other character is the presence of abundant spherules of cosmic origin, for these would almost certainly have become oxidised in old strata and could not therefore be detected. The only positive evidence available would be the extreme fineness of the component particles of the rock and the possible presence of zeolites, though these minerals also occur under other conditions.

CHAPTER XII

CLIMATIC BELTS IN THE OCEANS

Distribution. The belts of sedimentation described
in the former chapters occur concentrically around
the abyssal deposits of the central parts of the oceans,
with general parallelism to the coast-lines. They may
therefore trend in any direction along different parts
of their course. The climatic belts on the other hand
have a general east and west trend, with minor de-
viations due to the fact that they tend to run parallel
to the isotherms rather than to the lines of latitude.
They may therefore be parallel to the belts of sediment previously described, or oblique or even at right
angles to them, the amount of divergence depending
on the general trend of the coast-lines at any particular geological period, being least when the coasts
ran east and west. This will have important effects
on the distribution of organisms. In addition to the
previously described concentric belts, and to the
climatic belts which form the main subject of this
chapter, there are the marine life-provinces, which
add a further complication; but disregarding for the
moment these life-provinces, there is a general com-
munity of character of the faunas of any particular
climatic zone on a large scale, and this is of importance
to the geologist. The great climatic belts which seem
to be most useful are the Equatorial, the North and
South Temperate, and the North and South Frigid,

though it must not be assumed that these have always existed with temperatures similar to those now prevalent in each: study of the rocks of past times bears strong testimony to the contrary.

In studying the geological evidence as to climatic conditions both lithological characters and the nature of the included organisms afford valuable information. We may first consider lithological character.

Lithology. As a result of the study of lithological character one can hardly expect to distinguish the precise climatic belt in which any given deposit was formed, but merely to ascertain whether it was deposited under warm or cold conditions, and that only in the case of particular kinds of sediment. Cold conditions will be indicated by evidence of floating ice affecting the area in question during the formation of the sediments. Indications of floating ice are mainly furnished by the presence of boulders scattered through fine-grained deposits, for the fineness of grain indicates that the currents were but of moderate transporting power, and though the association of boulders and fine material may be found in some beach deposits, it does not occur far from land, except in the circumstances now being considered. The main cause is certainly floating ice, but there is one other that must be noticed, namely, transport by floating vegetation, in the form of single trees or matted masses forming "rafts" such as are carried down by many tropical and subtropical rivers, and often drift far out into the ocean before they ultimately become waterlogged and sink, carrying

M   14

down with them the stony burden they have borne from the land. Sometimes this mode of transport is indicated by carbonaceous matter surrounding the boulders but this will probably be but rarely detected except• when the boulders are embedded in light-coloured rocks, such as Chalk. When the boulders are actually transported by floating ice they will no doubt be most abundant in sea tracts not far removed from the area in which the ice was formed, so that at the present day the boulders are most abundantly deposited in the circumpolar regions, but will be deposited in ever-diminishing numbers far away from these, for we actually find icebergs as far north as the Cape (35° S.) in the southern hemisphere and as far south as the Azores (40° N.) in the northern hemisphere, so that only 75 degrees of latitude in the Atlantic are actually free from floating ice.

It may be observed that at present we know very little of the character of the marine boulder-bearing deposits, and practically nothing of such when laid down far from land. Capt. Feilden described some marine boulder-clays in Kolguev Island, but the most important examples of marine glacial accumulations are those described by I. C. Russell in the case of the Robinson, Chaix and Samovar Hills rising through the Malaspina Glacier as the result of faulting. These, however, must have been formed near the coast. They are incidentally very important as furnishing us with criteria whereby we may distinguish marine from terrestrial boulder-clays. Apart from their great thickness (4000 ft.), their most noteworthy features are the abundance of boulders, both angular and rounded, the well-marked stratification and the great quantity of marine shells contained in them[1]. As a matter of fact boulders such as might have been transported by ice are very rare in the fine-grained marine deposits formed in past times, though the exceptions when they do occur are of high importance. This is a subject to which further allusion will be made later.

Turning now to lithological evidence for the prevalence of high temperature conditions, the available evidence is mainly limited to special areas, such as have already been described in the cases of certain constricted seas of warm regions, for the probability is that definite criteria will be most convincingly applied where as the result of the occurrence of restricted waters and high temperature evaporation is carried on to some extent, and where there is an abundant supply of decomposing organic matter, the first favouring direct precipitation and the second bacteriological processes.

The cases which have been already examined, namely, precipitation of calcium carbonate, the formation of oolite, glauconite and phosphatic nodules, together with the study of their present geographical distribution in modern seas, suggest that at any rate they are formed most extensively in the shallows of warm seas, but the matter requires much further study before it can be definitely asserted that warm

[1] Russell, I. C., 13*th* *Ann. Rep. U.S. Geol. Surv.* 1893, p. 24.

waters are necessary for their formation on a large scale.

Organisms. Attention was previously drawn to the fact that in the case of the land it is sometimes dangerous to draw conclusions as to climatic conditions from the general character of the fauna. The same is the case when dealing with marine deposits. Geologists have been too prone to infer warm climatic conditions from the abundance and variety of fossils and the large size attained by many species. Baron A. E. Nordenskiöld drew attention to this in his account of the voyage of the *Vega*. The following extract from that book[1] refers to the results of a dredging off Cape Chelyuskin, the most northerly point of Asia.

"The yield of the trawling was extraordinarily abundant; large asterids, crinoids, sponges, holothuria, a gigantic sea-spider (Pycnogonid), masses of worms, crustacea, etc. It was the most abundant yield that the trawl-net at any one time brought up during the whole of our voyage round the coast of Asia, and this from the sea off the northern extremity of that continent....The temperature of the water was at the surface 0° to − 0°·6, at the bottom − 1°·4 to − 1°·6."

Nordenskiöld himself justly called attention to this occurrence as indicating the danger of inferring warm climatic conditions from the abundance and variety of a fauna which includes large forms.

[1] Nordenskiöld, Baron A. E., *The Voyage of the Vega*, London, 1881, vol. I, p. 350.

There are certain organisms however which probably justify us in assuming the prevalence of a warm climate at the time of deposition of the sediments containing them. Such are the reef-building corals. The importance of these may often be tested by observation of the organisms which accompany them. Dr G. Lindström (*On the Silurian Gastropoda and Pteropoda of Gotland*)[1] felt justified in inferring from the association of large *Trochi, Turbinidae* and *Pleurotomariae* with reef-building corals in the Silurian limestones of the island of Gotland that these deposits were laid down in warm waters. Similar associations occur in the Silurian limestones, and indeed in many deposits of various ages. It is interesting to note that these limestones often present some of those lithological features which have been mentioned as probably indicative of warm climatic conditions.

In the case of sediments of very ancient date there is however often considerable doubt as to the approximate temperatures at which they were formed, owing to the great difference between their organisms and those of more modern deposits. As we approach to recent times and the organisms resemble more closely those now existing the difficulties diminish, though except in the case of the fairly recent ones only a general impression as to the prevailing temperature can be gained. When the Pleistocene deposits are studied, where most of the species are still existing, estimations of the actual temperature may be

[1] Lindström, G., *Kongl. Svenska Vetenskaps-Akad., Handl.* vol. x, 1884, No. 6.

attempted. ·Forms now living in the Arctic waters, as certain species of *Yoldia*, give us definite evidence that during part of that period the Arctic province had shifted south as far as England, to recede again into its present position in Holocene times.

Climatic Belts in Past Times. Though it is difficult to make even an approximation to the actual temperatures of remote ages, attempts have been made to define the boundaries of different climatic belts in the distant past by study of marine faunas. It has been suggested that such climatic belts occurred as far back as Lower Palæozoic times. In Europe a very characteristic fauna is found in lands on either side of the Baltic and in the north of Scotland; a different one is found in southern Britain, and a third in central and southern Europe. At least two of these seem to be represented also in North America, the most northerly in Eastern Canada, and the middle one in New York. These variations may be due to other than climatic causes, and in any case the indications are too vague to carry any weight in the present state of our knowledge.

Another instance of asserted climatic belts of later date than that last mentioned is afforded by those originally defined by M. Neumayr in the case of the Jurassic and Cretaceous rocks. He divides the world into Equatorial, North and South Temperate and Boreal zones, that corresponding to the Boreal in the southern hemisphere being then unknown. The zones run generally parallel to the Equator, and are characterised by definite assemblages of fossils. The three zones of the northern hemisphere are all represented in Europe[1]. Neumayr's explanation of the distribution of the Jurassic and Cretaceous fossils is now generally discredited, the recorded facts being otherwise explicable. The possibility of tracing such zones does not however appear to be disproved.

Bearing of Wegener's Hypothesis. It is no part of this book to discuss Wegener's hypothesis, which is still in process of being tested. Suffice it to point out that should the theory prove to be true any original arrangement of climatic belts, whether parallel to the Equator or not, may be seriously altered as a result of continental drift, but such alteration need not necessarily occur, for drift in an easterly or westerly direction will leave the trend of normal climatic belts unaffected.

In connection with the subject of climatic belts in past times more than one point of interest arises.

Shifting of Belts. Whatever may be the cause we have actual evidence of the shifting of temperature zones, as the result of climatic change in space, as opposed to the drifting of land masses. Again and again we find that any particular area, such as England, has been subjected, sometimes to higher, sometimes to lower temperatures than now prevail. For instance, in the case of the British Isles there is evidence that, apart from epicycles, there was a general lowering of temperature from early Eocene to early Pleistocene. In addition to the time variation

[1] Neumayr, M., *Denksch. Math.-Naturwiss. Classe der k. k. Akad. Wiss.* vol. XLVII, 1883.

we also get proof of space variation, as for instance when tracing the nature of the organisms contained in the various strata from southern Europe to Greenland. In the case of each of the subdivisions of the Tertiary, however, although temperature diminishes as we proceed northwards, the organisms mark higher temperatures than those now prevailing. It would appear then that the Tertiary climatic belts, though probably sub-parallel to the present ones, had higher average temperatures throughout.

This appears to have been the general rule not only in the Tertiary and for a limited area, but from the time of the first-formed fossiliferous rocks to modern times, and affecting vast regions of the two hemispheres, the exceptions being only temporary and infrequent. Baron Nordenskiöld long ago expressed a somewhat similar opinion, pointing out that in the Arctic regions fossiliferous deposits of various ages from Lower Palæozoic to Recent exist, and that nearly every great geological system is represented: nevertheless no indication of ice action occurs until Pleistocene times. It is true that evidence of a very early glaciation has been discovered within the Arctic Circle, but this does not seriously affect the cogency of the argument. Dr Coleman, in his recent book *Ice Ages*, definitely regards glacial periods as exceptional, occurring only at widely distant intervals, with generally warm conditions prevailing the world over in the intervening periods, which were of much greater duration. If this be the case, climatic belts, though no doubt existing, would almost certainly not be marked by such great differences of average temperature as at present exist.

An interesting subject that concerns geologists is that of bipolarity—the occurrence of relationships between the faunas of corresponding belts in each hemisphere, which are not shared by those of the intervening belt or belts. It is doubtful whether geology can throw much light on this interesting question at present, though no doubt it will furnish its contribution later, when the distribution of fossils has been studied in much greater detail.

Crossing of Belts of Sediment and of Climate. At the present day the climatic belts have an east and west trend, while the belts of sediment may run in any direction, though the prevalent trends are approximately north and south, and in a less degree east and west. Thus the general direction of the coasts bordering the Atlantic and Pacific Oceans is north and south, those of the Arctic Ocean and the north of the Indian Ocean east and west. It is clear that the coincidence or otherwise of the belts of sediment and of climate must exercise an important influence on the distribution of organisms. Planktonic forms will no doubt be affected to a less degree than benthonic forms, owing to their rapid and wide dispersal, living or dead. Of the benthonic organisms those of the zones farthest away from the coasts will no doubt be affected in a less degree than those nearer the shore: so far as temperature conditions are concerned, the greater the depth the less will the temperature be affected by solar heat, and the range of temperature with latitude will be less in deep water than in the shallows of the belt of variables.

The organisms in the belt of variables will be affected in a marked degree. Where the belt runs more or less at right angles to the climatic belts dispersal will tend to be slow and only to extend for short distances, whereas when the two kinds of belt coincide there may be little to check dispersal along the littoral, which may then proceed for long distances and affect a large number of species. Under these conditions then such wide lateral distribution as aids correlation will be most effective, and we may find the same fauna distributed over very wide areas during a brief geological period.

At the present day it would seem that coincidence of climatic and sedimentary belts is comparatively limited: may it not have been more general in past ages? Reference has already been made to the predominance of an east and west sea between Gondwanaland and Angaraland in Mesozoic and early Tertiary times, and its possible existence in Palæozoic times. When it existed, conditions would be favourable for an east and west spread of its organisms, including those of the belt of variables, if the climatic belts of the period also ran east and west, which as we have seen cannot be inferred with certainty.

In Palæozoic times there was a remarkable resemblance between certain contemporaneous faunas, such as those of the Devonian of N.W. Europe and of Canada, these belonging largely to the belt of variables. Now the physical evidence is in favour of a land to the north, whose coast had on the whole a trend in an easterly and westerly direction. May not this have favoured dispersal of the shallow-water fauna to a marked degree, thus bringing about the similarity of organisms actually occurring? It may be that this similarity of faunas, and other instances that might be quoted as found elsewhere, do not necessitate any continental drift separating once adjacent land masses.

A detailed study of the conditions prevalent in belts of sedimentation and in climatic belts, and of their combined effects, will no doubt throw much further light upon the distribution of organisms in past times. Much has already been done, and we await with confidence still greater advances in this direction.

There are still many apparent anomalies in the distribution of organisms in past times which are still unexplained. Physical barriers are usually invoked in such cases, but many of these barriers are postulated for want of a better explanation, and in many cases the existence of such hypothetical barriers has been subsequently disproved. I have said little about land barriers, which will, like different climatic belts, cause very different faunas to occur in strata laid down at no great distance from one another. It is obvious that detailed study of the geology of the region must be made in order to show which cause was in operation. In either case although two contemporaneous faunas differ some species will be common to the two areas. A land barrier must no doubt act as a deterrent to the spread of many organisms, but there is evidence that organisms can work round the seaward termination of a barrier, thus reaching one partly constricted water tract from another.

There is one subject not yet touched upon, namely, the possible generally wider distribution of organisms in past times. This will be considered in the next chapter. From what has been said it will be concluded that the study of the distribution of life in the past is still in its infancy, and that it affords a fertile field for future geological workers.

CHAPTER XIII

UNIFORMITY AND EVOLUTION

THE Huttonian principle of uniformitarianism has done good service in the past, and so far as the rocks from Palæozoic times onwards are concerned is far from being valueless; indeed a modified form of this principle is still of very great service. The Huttonians regarded the origin of the earth as being outside the province of the geologist, but at the present day it is regarded as within his domain, and if this be so, it is pretty clear that the principle of uniformity will here fail. But when dealing with such events as occurred before the formation of the oldest known rocks, the geologist cannot rely on observation, but only upon information furnished by followers of other sciences, and in that sense Hutton was right in ruling these early events as outside the scope of geology. At the time when matters had progressed further and actual records of what happened can be obtained, based on study in the field, general conditions were beginning to be similar to those now prevalent, and a modified uniformitarianism then becomes of use. This will be discussed in the present chapter.

It must not be forgotten that only a very small part of the earth's mass is available for direct study by ordinary geological methods. The thickness of the sedimentary rocks has been variously estimated at from 50,000 to 100,000 ft. Even the lower estimate may be excessive, but probably these figures give some indication of the order of magnitude. If we take the higher figure the rocks with which we can deal directly in attempting an elucidation of earth history form only $\frac{1}{200}$ of the earth's radius. The geologist is in fact in the position of a dermatologist who from a study of the human skin endeavours to work out the structure and development of the whole body.

Of what lies below this skin we know practically nothing. It may no doubt have been formed in a comparatively short space of time, whatever may have been the earth's origin. Even when we enter upon the study of the available evidence we are confronted with a great thickness of almost unfossiliferous rocks of Pre-Cambrian age, where the story is still fragmentary, and the validity of uniformitarianism as applied to these rocks is difficult to estimate, though we seem to have indications that it must be applied in a manner far from strict. The Pre-Cambrian rocks probably represent more than half the thickness of the sedimentary rocks preserved to us, and as regards their origin much still remains to be learnt, compared with what has already been ascertained concerning the rocks yielding abundant fossils from Cambrian times onwards. During Pre-Cambrian times much may have happened which cannot be accounted for if we adhere too closely to uniformitarian principles.

We are concerned in this book only with the operation of the agents which gave rise to the formation of the sediments. The working of these is dependent upon the heat received from the sun, that of the earth's interior and gravity. In the present state of our knowledge the idea that the heat received from the sun and that of the earth's interior have been diminishing throughout geological times seems no longer to be a certainty. We are free therefore to draw our own conclusions from what we observe by study carried out on geological lines unfettered by limitations formerly imposed on us by physicists.

A comparison of the lithology of the most ancient sediments with those now being formed does apparently indicate that apart from minor details the operation of the agents responsible for their production was indeed generally similar, both in kind and in intensity, to that concerned in the making of the modern sediments, so that even in Pre-Cambrian times uniformitarianism in a wide sense forms a useful basis.

A cursory examination of the general character of the sediments shows this, for the similarity of the various pebble-beds, sands, muds, and other deposits to those now being formed, obviously indicates generally similar conditions, but similarity of conditions under which agents worked is specially brought out by study of those of climate, which shows that the temperature changes, so important as controlling factors in the production of sediments, were even in Pre-Cambrian times very similar to those which have occurred in modern ages. When we find evidence of glacial action in rocks it is clear that temperature conditions must have been on the whole like those now prevalent, and that heat received from the interior of the earth cannot have been a serious factor in producing the temperature conditions, nor can the amount of heat received from the sun have been greatly in excess of what it now is. We meet with the earliest glacial accumulations in Pre-Cambrian times, and they are repeated at intervals later, so that from an unknown Pre-Cambrian date to the present day the principal agents concerned in sedimentation were working on the whole as now.

The question is, when did fairly uniform conditions become established? We agree that the origin of the earth is not to be accounted for on uniformitarian lines, and probably agents were different both in kind and in intensity from what they now are for an unknown length of time after the earth was actually formed. We desire to know how long a time elapsed before general uniformity began to prevail, compared with the subsequent ages during which it has prevailed. Let us for convenience regard the latter as beginning in Cambrian times, though as pointed out it probably began earlier. From the complex and highly organised fauna found in the lowest Cambrian rocks, Huxley argued that as much time passed by before Cambrian times as subsequently in order to allow of its evolution. This is probably an underestimate and it is quite possible that if we could divide the earth's history into say twenty volumes of roughly equal size, ten of them might be concerned with what happened before our earliest known rocks were formed, and five more with the formation of those Pre-Cambrian rocks of which we have knowledge, though comparatively slight, leaving only five to deal with events from Cambrian times onwards.

These figures are of course purely conjectural, but it is important that the geologist should realise that he is dealing at first hand with only the latest part of the earth's history. The case is comparable with that of the student of human history. The origin of man, as of the earth, is at present a matter for speculation: the history of primæval man is still very obscure, and it is only when we reach historic times, forming a small fraction of the total period of man's existence, that our knowledge becomes detailed and precise.

In dealing with the sediments during times when general uniformity on a large scale had set in, very important departures from such uniformity may have taken place, and these may now be considered.

Lands and Oceans. Even if the agents of erosion were potentially more effective in earlier times than at the present day, their actual effects might be less, if the distribution and character of land and sea were different then and now: for, however potent the agents might be, nothing would be done in the way of sub-aerial erosion if there were no lands to be eroded, and with comparatively small land masses of little altitude erosion would be slight. The theory of the growth of continents by accretion, which receives some support, does not necessarily indicate smaller continents in past times, for destruction might proceed *pari passu* with accretion, but the possibility of smaller land areas must be taken into account. It would affect the quantity of eroded material rather than its character.

Again, the oceans may have had a smaller depth

than those now existing. This would cause a greater quantity of shallow-water sediment to be laid down, but at present we have little or no information upon this subject; it is one which obviously requires attention.

The nature of the coast-lines would have important effects. These may have been straighter or more indented than those of modern times: in the first case the belts of sediment would run more regularly than now and the minor variations would be less complex; in the second case the contrary would occur. Here again we must await the results of more detailed observations.

Primitive and Later Sedimentation. Whatever view may ultimately be adopted to account for the earth's origin, it is generally agreed that the rocks of the primitive crust were of an igneous nature. Accordingly the first-formed sedimentary or derivative rocks would retain much of the soluble material supplied by the eroded crust, and this would be aided by processes to be presently noted. The first-formed sediments then would contain a large amount of alkalis and alkaline earths, iron and magnesia. As time proceeded the already formed sediments acted on by agents of erosion would themselves supply material for fresh sedimentation and soluble material would be further leached out from them, but in the meantime a considerable part of the crust would still consist of igneous rocks. At later periods the amount of igneous material exposed would diminish and that of sedimentary rock increase, and as ancient sediments were again and again exposed to erosion the amount of soluble material which they contain would be ever growing smaller in amount. It would appear, there-fore, that in the normal course of events the mechani-cal sediments formed during the first erosion of the primitive crust would, other things being equal, con-tain the largest proportion of soluble material, and those formed at the present day the smallest. In the modern sediments, therefore, we should expect to find the greatest proportion of silica and aluminium silicates.

Influence of Terrestrial Vegetation. It appears to be now established that terrestrial vegetation did not get a definite footing before Devonian times. It has been pointed out that this would have an effect upon erosion. Firstly, the absence of soil would cause the rainfall to run off rapidly, and thus accelerate mecha-nical transport and increase its effect. The conditions would be to some extent comparable with those of the violent floods which often occur in desert regions. The absence of soil would also have a restraining influence on chemical solution, for such solution would be brought about only by acids derived from the air, and not by those due to the decay of vegetation.

Dissolved Salts in Sea-water. Prof. Joly has ad-vanced a view that the waters of the primitive oceans were fresh and that the salinity of the modern oceans has been attained by gradual introduction of the soluble matter brought down by rivers, being in fact the excess over that reconverted to the solid

condition by chemical or organic agencies. This speculation has a certain amount of bearing on our present
subject, as for instance in producing the flocculation
of mud particles in the presence of soluble salts. More
important is the presence of certain soluble matters
as affecting organisms. It has already been argued
that certain salts are inimical to the growth of organisms which secrete calcareous tests, and it was suggested that compounds of iron in particular have this
effect. The supply of such compounds in the earlier
oceans may have been very considerable, for more
than one reason, but especially on account of the
great amount of igneous and metamorphic rocks
then undergoing erosion. To this matter we shall
recur.

Marine Vegetation. The abundant occurrence of
black mud, often containing graptolites, in the Lower
Palæozoic rocks has been noticed in a previous chapter.
It was pointed out that some of the colour is due to
carbon, attributed by Lapworth to the decay of algae.
It is possible that algae then played a more important
part in affecting the nature of sediments than is the
case at the present time, but there appears to be little
evidence bearing on this point.

Effect of Vulcanicity on Sedimentation. It has been
noted that in addition to material supplied by erosion,
varying amounts of volcanic material contribute to
the formation of sediments at the present time. May
not this have been the case to a greater extent
formerly? Examination of the Pre-Cambrian sediments does indicate a very great amount of such

material, and there is also much in the Palæozoic rocks of Britain. Observations in detail have not been made over a sufficiently wide area to justify the inference that vulcanicity did play a more important part in earlier times than now as a contributor to sedimentation, and even should it prove to be the case it does not follow that the decreasing importance of volcanic material was due to a steady falling off of the supply as times approached nearer to the present day. The observed phenomena might be produced by recurrence, of which more anon.

Distribution of Organisms, Vertical and Lateral. Little need be said of the evolution of organisms from the simple to the complex, for at the time when the strata first furnish us with abundant fossils, namely, in the early Cambrian beds, evolution had already progressed so far that all the phyla of the invertebrates had come into existence; only the vertebrates made their appearance later. A question that has puzzled all geologists is the sudden appearance of the varied and complex fauna of the Cambrian period. This is a subject which has long baffled them, and at one time an attempt was actually made to cut the Gordian knot by imagining the sudden appearance of varied forms of life on our planet as due to transport from some extra-terrestrial region. It is, however, regarded as certain that evolution to the state of complexity manifested by the earliest Cambrian fauna did take place on the earth itself, and notwithstanding the obscurity of the forms discovered in Pre-Cambrian strata, it is now generally admitted that we are

cognisant of relics of organisms belonging to those far-
distant ages. The difficulty is to account for their
rarity. The question was considered in a series of
papers read before the International Geological Congress at Stockholm in 1910 and published in the
Report[1].

More than one suggestion has been made and probably no one explanation is sufficient in itself. There
is a certain amount of evidence pointing to a non-
marine origin of many of the Pre-Cambrian strata,
and it has been suggested that in Pre-Cambrian times
the marine tracts were on the sites of our present deep
oceans, in which case the Pre-Cambrian marine fossils
must be concealed below the present ocean floors.
This may have contributed to the scarcity of organic
remains, but is probably not the main cause. A more
probable explanation is that the earlier organisms
did not secrete calcareous tests, at any rate to so great
an extent as the later ones. This might be brought
about by a widespread occurrence of those conditions
which have been shown to be inimical to the existence
of such, as in the case of the Old Red Sandstone,
the *Posidonomya* beds of South Wales, and the graptolitic mudstones.

Or, again, organisms may have been evolved to a
state of high complexity before the power of secreting
calcareous matter was attained. In early times, when
life was comparatively simple, the struggle for existence would not have reached its full intensity and
protective coverings would not be so necessary. As

[1] *C.R. XI Congr. Géol. Internat., Stockholm,* 1910, p. 105.

it is, calcareous secretions are no sign of high organisation at the present day. In even the highest phylum of the invertebrates, the Mollusca, forms are found which do not secrete calcareous shells, and we find in other phyla closely allied groups, some secreting hard parts, others not, as witness the corals and sea-anemones. Nor is this possibility based entirely on imagination. Of the recognised Pre-Cambrian organisms, some, the Annelids, are represented by casts of soft-bodied forms; others, like *Beltina*, secreted what was apparently a chitinous test, devoid of lime. An interesting occurrence is found in the oldest Cambrian rocks among the trilobites. The *Olenellus* group contains a number of genera terminating with a long spine, formerly regarded as a pygidium, though now known not to be such, but the spine of the fifteenth thoracic segment. In the genus *Mesonacis* this spine also occurs on the same segment, but behind it are 7 or 8 further thoracic segments of very tenuous character, apparently membranous, before the pygidium is reached. It is almost certain that the same occurred in *Olenellus* itself, but that the terminal thoracic segments were of such extreme tenuity that they have perished, or at any rate have not hitherto been detected.

That a large number of organisms of high complexity lived in Cambrian times and did not secrete hard parts, or only chitinous ones, is amply proved by their abundance when circumstances were favourable to their preservation, as for instance in the very fine-grained Burgess Shale of the Middle Cambrian of

British Columbia. There seems, therefore, to be nothing improbable in the existence of organisms for long periods of Pre-Cambrian time without secretion of hard parts, in which case their remains would only be discoverable under exceptional circumstances. When we consider how rare is the occurrence of such faunas as that of the Burgess Shale we need not despair of eventually finding such a fauna in the Pre-Cambrian rocks.

It was inferred that a great lapse of time occurred between the first appearance of life and the Cambrian period, and this inference was based upon the complexity of the Cambrian fauna, but we may be prone to over-estimate it, for biologists have suggested that evolution proceeded at an exceptionally rapid rate in early times.

Turning now to distribution of organisms in space as bearing upon similarity or otherwise of conditions in early and late geological times, one matter of interest deserves special notice. Pictet enunciated a series of palæontological laws, of which No. 9 was as follows: "The geographical distribution of species in earlier periods was more extended than the range of species of existing forms." Highly probable as this may be it can hardly be regarded as attaining to the dignity of a law. After careful consideration, there does appear to be evidence in its favour, and on theoretical grounds it seems to be probable, for if conditions in the inorganic world were simpler than now, and the life of earlier times was simpler and less varied than later, allowing of dispersal of organisms unchecked by so many barriers, inorganic and organic, as existed later, these organisms should spread laterally to a greater extent than they did subsequently. But it would require much more work in comparing ancient and modern faunas before the law could be regarded as established. One serious difficulty is due to the tendency to split genera into a greater number of species in the case of comparatively recent organisms, so closely related to existing forms, than when dealing with ancient forms often belonging to entirely extinct groups. This in itself would vitiate the statistics, making the lateral distribution of more modern forms appear more restricted than that of the ancient ones. The subject, therefore, requires much further study than has up to the present been devoted to it.

Recurrences. Apart from the question of increasing complexity in the inorganic and organic worlds as the result of evolution, it is quite clear that the changes have not always been steadily in one direction, as implied by the principle of uniformitarianism. Recurrences had considerable influence in modifying such steady advance. From the point of view of influence upon deposition, recurrences of much importance are those connected with earth-movement, vulcanicity and climate. In each of these cycles have occurred occupying long periods of time, with epicycles of various degrees of magnitude.

The important effect of cycles of earth-movement from the present point of view is the alternation of periods of positive and negative changes, producing alternate submergence and emergence, thus giving rise to alternations of sea and land over the critical areas more extensive than the average. During the periods of submergence the deposits of the great geological systems were laid down; during those of emergence parts of them were eroded. Each geological system would have ideally its three phases: the early one of shallow water, of restricted distribution, the middle one of deeper and more extensive waters, and the last of shallowing and retreating waters. This ideal simplicity seems always to have been complicated by epicycles of similar changes on a smaller scale. The tendency of recent work seems to show to an increasing degree that the great revolutions had to some extent world-wide effects. As bearing on this question reference may again be made to Prof. Chamberlin's essay on "Diastrophism as the Ultimate Basis of Correlation[1]," already referred to. These revolutions had most important effects on sedimentation and the characters of the sediments. How many major revolutions occurred is a matter of opinion, and opinions differ, but they may probably be counted on the fingers of two hands.

That there were cycles of alternate activity and quiescence in volcanic action is clear, though it is far from clear that these manifestations were world-wide. It was once assumed that they were, but as our knowledge of the geology of distant regions increases,

[1] In Willis, Bailey and Salisbury, R. D., *Outlines of Geologic History*, Chicago, 1910.

grave doubt has been thrown upon this. That being
so, it will be safer to confine our attention to those
regions where the geology is best known in the northern hemisphere. Activity in Pre-Cambrian times
seems to have been exceptionally great, though no
doubt not continuous. In Palæozoic times also
activity was marked, though here undoubtedly interrupted by epicycles of quiescence. In Mesozoic times,
on the contrary, the evidence seems to indicate
quiescence, of cyclical importance, to be replaced in
Tertiary times by violent activity, initiating a new
cycle in which we are probably still living.

Climatic cycles also seem to be well established.
On the whole it would appear that temperatures were
generally higher in the past than they are at present.
We are living in a cycle of low temperature. Such
cycles have occurred more than once in past times.
At least one happened in Pre-Cambrian times, another
in the Cambrian, another during the Permo-Carboniferous, another probably in the Cretaceous, and yet
another in the Pleistocene and Holocene, of which an
epicycle is now with us, marked by somewhat higher
temperatures than the mean of the cycle. Prof. Coleman (*Ice Ages*) is apparently convinced of the comparative shortness of the cold cycles and the greater
geniality as a whole of the climates of the past as
compared with those of the present.

It would appear from all considerations that we are
living in an epicycle of a cycle of emergence in one of
the great revolutions; in an epicycle of a cycle of
vulcanicity; and in an epicycle of a cycle of low

temperature. In no one of the three are we probably at the maximum of the cycle, which, so far as we know, as regards earth-movement was in Miocene times, with regard to vulcanicity in the early Tertiary, and with regard to climate in the Pleistocene. In any one of these it is possible that the maximum is not yet past, but to come. For instance, we may be living in an interglacial epicycle of a glacial cycle still incomplete, and possibly still to attain its greatest intensity in the future. It is generally conceded that earth-movement and vulcanicity are genetically related, and it is possible that there is also a genetic connection between earth-movement and climate. Several lines of evidence seem to point to this, and what we have said concerning the coincidence of cycles of earth-movement, vulcanicity and cold in those geological times in which we are now living affords some support.

It is clear that recurrence must be of very great importance in affecting the uniformitarian doctrine of the operation of agents in past times in the same manner and with the same intensity as now; of an importance possibly at least as great as the supposed gradual slowing down of these agencies from early times to the present day. For if the present period be exceptional its phenomena do not furnish a fair standard for comparison with those of the past. The tendency to regard the present as normal has led us into several difficulties, and this is in itself a warning against a too rigid adherence to uniformitarianism. A sure path is that of evolutionism, from the simple to the more complex, with oscillations now on one side, now on the other, of a smooth curve which might be taken to indicate the gradual change from simplicity to complexity.

Conclusion. An attempt has been made in the foregoing pages to give an account of the main features of the sedimentary deposits which remain as they were formed, apart from certain changes which may be briefly noticed. Of such changes as are included under the heading metamorphism it is unnecessary to speak. But minor changes have occurred, chiefly induced by pressure, by cementation and by crystallisation. The effect of pressure has been to compact incoherent material. This has also been brought about by cementation, the loose particles having been bound together by such substances as silica, calcium carbonate and iron compounds. Crystallisation has converted certain originally amorphous limestones into crystalline rocks. These changes are mentioned because they may mask the similar origin of superficially dissimilar deposits. For example, it has been mentioned that certain now indurated beds of Carboniferous limestone probably had an origin similar to that of part of the Chalk, from which they now differ greatly in appearance.

It has been my object to show that, apart from differences in detail, conditions spread over extensive regions have produced wide-spreading belts of sediment, similar in lithology and organic contents. It would appear that climatic belts exercise control of the deposits and their organic contents over both land and sea. In the sea tracts, however, these climatic belts are of minor import when compared with the lithological belts arranged concentrically around the central ocean depths, and trending parallel with the coast-lines. These belts have advanced towards and over, and retreated away from, temporary land areas during alternate periods of positive and negative movement, causing overlap of more open-water belts upon the shallower ones, and vice versa. If the theory of the permanence of ocean basins be even approximately true, the changes noted will recur again and again in the same area, the sediments in the landward parts of the oscillating water tract being separated by unconformities. Sometimes transgressions will take place to a greater extent than at others, causing the waters to encroach more towards the centres of the land masses; gulfs and other epicontinental waters will occupy the lower tracts of these land masses during periods of submergence, and in many regions these epicontinental waters will spread over them time after time.

I have endeavoured to show that N.W. Europe, including the British Isles, is exceptional in that these epicontinental waters have again and again occupied parts of it: the area has seldom, if ever, formed part of a real open ocean. I have further argued that the present times are exceptional. It follows, therefore, that neither the Holocene period in time, nor N.W. Europe in space, can be used as standards for comparison when dealing with the sedimentation

of the past. The geologist must not be too insular;
he must collect his data from past and present alike,
and seek it, not locally but the world over. A vast
amount of fresh information has still to be collected;
the study of sedimentation is as yet in its initial
stage; further progress will be made by geologist
and geographer alike, probably in fullest measure
by those who are acquainted with both sciences.

INDEX

Abyssal deposits, 198
Aeglina, 36
Agulhas Bank, 117
Algae, 24
Ammonites, 23
Ancylus Lake, 171
Angaraland, 218
Angelin, N. P., 13
Aralo-Caspian basin, 67, 157
Arber, E. A. N., 142
Archæopneustes, 205
Archanodon, 45
Arethusina, 177
Aridity, 63
Ash, 124
Atolls, 126, 131

Bacteria, 118, 161, 188
Bahamas, 160
Bailey, E. B., 120, 164, 195
Baltic Sea, 154, 171
Barbados, 205
Barrande, J., 21, 35
Barrell, J., 166
Barriers, 23, 27, 219
Bateson, W., 67
Beltina, 231
Belt of Variables, 75, 89
Belts of altitude, 48
Belts of sedimentation, 73
Benthos, 23, 25, 107, 134, 176, 183, 217
Betula nana, 57
Black Earth, 59
Black mud, 182
Black Sea, 183, 188
Blue mud, 174
Bonney, T. G., 181
Boulder-clay, 52

Boulders, 56, 210
Brachiopods, 133
Braided channels, 51
Branchipus, 171
Burgess Shale, 232

Cambridge Greensand, 37, 93
Cenomanian transgression, 11
Ceratopyge, 14
Chamberlin, T. C., 11, 234
Chasmops, 33
Chemical precipitates, 40, 66, 107, 129, 156, 161, 164
Chert, 204
Chronology, 3
Classification, 5, 7
Climatic belts, 48, 208, 219
Coal, 141
Coastal plain, 83
Cockle, 27
Coit, G. E., 102
Coleman, A. P., 216, 235
Colonies, 21
Columella, 57
Conglomerates, 106, 147, 166
Conocoryphe, 14
Continental shelf, 77, 83, 89, 106, 125
Contortion, 52, 56
Conularia, 170
Coralline zone, 100
Coral reefs, 109, 125, 131, 213
Coral sands, 125
Crinoids, 133
Cromer Forest Bed, 39, 42
Currents, 91, 120
Cyclas, 61
Cycles, 8, 70, 82, 233
Cystechinus, 205

M

Printed in the United States
By Bookmasters